GIS Tutorial for
Crime Analysis

Wilpen L. Gorr
and Kristen S. Kurland

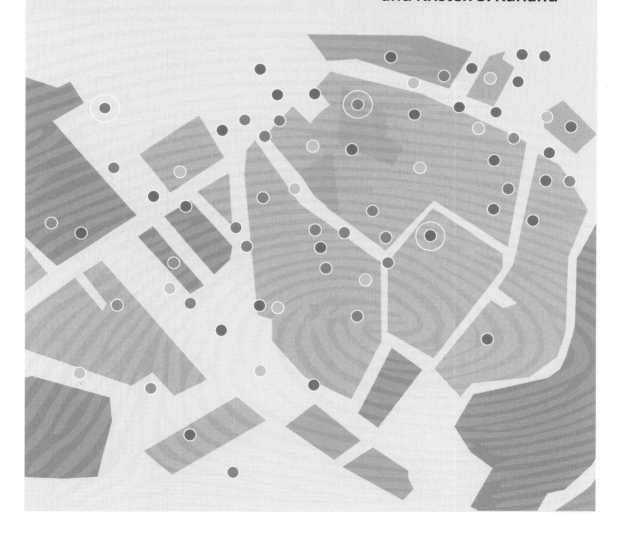

Esri Press
REDLANDS|CALIFORNIA

Esri Press, 380 New York Street, Redlands, California 92373-8100

Ask for Esri Press titles at your local bookstore or order by calling 800-447-9778, or shop online at www.esri.com/esripress. Outside the United States, contact your local Esri distributor or shop online at www.eurospanbookstore.com/esri.

Esri Press titles are distributed to the trade by the following:

In North America:

Ingram Publisher Services
Toll-free telephone: 800-648-3104
Toll-free fax: 800-838-1149
E-mail: customerservice@ingrampublisherservices.com

In the United Kingdom, Europe, Middle East and Africa, Asia, and Australia:

Eurospan Group
3 Henrietta Street
London WC2E 8LU
United Kingdom
Telephone: 44(0) 1767 604972
Fax: 44(0) 1767 601640
E-mail: eurospan@turpin-distribution.com

Contents

Preface   *ix*
Acknowledgments   *xv*
Learning pathways   *xvii*

Part 1   **Working with crime maps**

Chapter 1   ***Introduction to crime mapping and analysis   1***

Learning the system   *2*
Mapping and analyzing crime   *4*
Environmental criminology references   *7*

Chapter 2   ***Exploring ArcGIS Desktop   9***

Overview of ArcGIS Desktop software   *10*
Learning about ArcGIS   *11*
Tutorial 2-1   Exploring the ArcMap user interface   *12*
Tutorial 2-2   Exploring the ArcCatalog user interface   *18*
Tutorial 2-3   Examining map layer properties   *22*
Tutorial 2-4   Examining Layout View   *24*
Assignment 2-1   Critique an online crime mapping system   *26*
Assignment 2-2   Compare crime maps for serious violent crimes
in Pittsburgh   *27*

Chapter 3   ***Using crime maps   29***

Crime maps and their uses   *30*
Using crime mapping and analysis system maps   *33*
Tutorial 3-1   Using maps designed for the public   *34*
Tutorial 3-2   Using an early-warning system map   *38*
Tutorial 3-3   Using a pin map for field officers   *46*
Assignment 3-1   Analyze hot spots for larceny crimes   *51*
Assignment 3-2   Create maps for the media   *53*

Chapter 4   ***Designing and building crime maps   55***

Map design   *56*
Building crime maps   *58*
Tutorial 4-1   Building a pin map for field officers   *59*
Tutorial 4-2   Building an early-warning system for investigators   *69*
Tutorial 4-3   Building a map for public use   *78*
Assignment 4-1   Build an auto theft pin map   *82*
Assignment 4-2   Build auto squad choropleth maps   *85*

Part 2   Conducting crime analysis

Chapter 5   ***Querying crime maps   87***

Crime query concepts   *88*
Building and using crime queries   *90*
Tutorial 5-1   Creating attribute queries   *91*
Tutorial 5-2   Creating spatial queries   *98*
Assignment 5-1   Analyze leading-indicator crimes by day vs. by night   *102*
Assignment 5-2   Analyze robberies near check-cashing businesses   *104*

Chapter 6   ***Building crime map animations   107***

Crime map animation concepts   *108*
Building crime map animations   *108*
Tutorial 6-1   Building an animation for serial crimes   *109*
Tutorial 6-2   Building an animation for hot spots   *115*
Assignment 6-1   Build an animation for hot spots of illicit drug dealing   *122*
Assignment 6-2   Build an animation for leading indicators of
serial crime   *124*

Chapter 7   ***Conducting hot spot analysis   127***

Hot spot analysis concepts   *128*
Conducting hot spot modeling   *130*
Tutorial 7-1   Testing for crime spatial clusters   *131*
Tutorial 7-2   Using kernel density smoothing   *135*
Tutorial 7-3   Conducting Getis-Ord Gi* test for hot spot analysis   *140*
Assignment 7-1   Assess impact of date duration on hot spots   *146*
Assignment 7-2   Examine broken-windows theory for simple and
aggravated assaults   *148*

Part 3   **Building a crime mapping and analysis system**

Chapter 8   *Assembling jurisdiction maps   151*

Basemaps and jurisdiction maps   *152*
Assembling police jurisdiction map layers   *153*
Tutorial 8-1   Downloading and preprocessing basemaps   *154*
Tutorial 8-2   Extracting jurisdiction maps   *161*
Tutorial 8-3   Joining census data to polygon maps   *166*
Tutorial 8-4   Creating new map layers from basemaps   *168*
Tutorial 8-5   Digitizing features   *174*
Assignment 8-1   Download and use a census block group basemap
                            and data   *182*
Assignment 8-2   Create maps for foot patrols and DUI target areas   *184*

Chapter 9   *Preparing incident data for mapping   187*

Data processing and geocoding concepts   *188*
Geocoding crime incident data   *190*
Tutorial 9-1   Address matching, or geocoding, data   *191*
Tutorial 9-2   Improving address matching results   *195*
Tutorial 9-3   Processing update and master data files   *205*
Tutorial 9-4   Aggregating data   *209*
Tutorial 9-5   Protecting privacy in location data   *212*
Assignment 9-1   Geocode Pittsburgh 911 calls for service data   *217*
Assignment 9-2   Build space and time series data for 911 calls   *220*

Chapter 10   *Automating crime maps   223*

ModelBuilder concepts   *224*
Using ModelBuilder for automation   *225*
Tutorial 10-1   Exploring a completed model   *226*
Tutorial 10-2   Processing police reports into master files   *231*
Tutorial 10-3   Producing a pin map for field officers   *238*
Assignment 10-1   Build a model to produce choropleth maps   *246*
Assignment 10-2   Build a model to produce size-graduated point
                            marker maps   *248*

Appendix A   Tools and buttons   *251*
Appendix B   Task index   *255*
Appendix C   Handling data and homework files   *261*
Appendix D   Data source credits   *263*
Appendix E   Data license agreement   *265*
Appendix F   Installing the data and software   *269*

Preface

GIS Tutorial for Crime Analysis is designed to show how GIS technology can be used to conduct the type of crime mapping and analysis done by police officers every day. In addition, the book advances the field by including state-of-the-art and innovative approaches and tools.

Purpose and audience

The purpose of this computer workbook is to teach crime mapping and analysis skills using ArcGIS Desktop software. The book, which provides tutorials on the use of ArcGIS, is intended for self-learners as well as students in the classroom. A lot of it entails "learning by doing," and so, the book includes step-by-step exercises and brief written introductions to underlying criminology theories and crime analysis methods. It also includes references and Internet search terms for acquiring additional knowledge, which should be helpful for self-learners and motivated students.

GIS Tutorial for Crime Analysis includes much of what you need to know to become a successful crime mapper and analyst. You get the basics used in the field, plus the latest in state-of-the-art methodology. Most importantly, this book is not just about making crime maps and conducting crime analysis studies. The central theme of the book is that you must be able to build, maintain, and use a complete crime mapping and analysis *system* to fully meet the information needs of a police department and its many user groups. By building, maintaining, and using this system, you can take the daily flow of police reports from the field as input, perform comprehensive processing of the data by using GIS technology, and then output all the varied maps and analysis needed by field officers, investigators, police executives, and the public. You can even automate the system for increased efficiency by using the macros covered in the final chapter.

For the self-learner and classroom student

Besides learning how to build and use crime mapping and analysis systems with the use of this book, you will learn the underlying concepts and principles of crime mapping and GIS by doing the comprehensive exercises. Each chapter begins with a few pages of introduction, and then leads you step by step on your computer in the best practices of working with ArcGIS. The book includes all the critical screen captures showing ArcGIS processes at work so you can figure out what you are supposed to be doing, see the intended output, and have these materials at your disposal for study or review. If you get lost or make mistakes, you can look at the solutions we have filed for all exercise steps in a computer folder called FinishedExercises, which is on the Maps and Data DVD that comes with this book.

Although you can get through the step-by-step exercises quickly, there are times when you will need to slow down and internalize the workings of ArcGIS to make it your own. Working through the exercises makes you aware of how GIS works, but it does not complete the learning process. This book offers two features especially designed to help you learn more about GIS. First are the Your Turn assignments, interspersed throughout the chapters, which you need to complete to learn the material but do not need to turn in for grading. In a Your Turn, you repeat the steps just taken in an exercise, but do it on your own for a slightly different scenario. If you do not remember the steps, you can look back at the exercise as a reminder. Your Turn work does not take long, and the outputs are often needed for the steps that follow, so be sure to do them all. The second and most effective way to learn GIS is to do the challenging, end-of-chapter assignments, which require independent thinking. Doing the assignments takes time, but learning how to do them is your best bet for becoming proficient at using GIS. If you are a student in a class, it is important to learn how and where to save assignment files so your instructor can grade them when you submit them (please see appendix C). The final step is to do a crime mapping project on your own. Doing such a project will involve integrating and applying the knowledge and skills you have learned throughout the book. If you have forgotten how to use certain toolbars or how to do tasks or steps while working on a project, you can look them up in appendixes A and B. For a project, you can use the crime data included on the Maps and Data DVD that comes with this book or obtain crime data from a police department. A good place to get crime data on the Internet is from the Washington, D.C., Metropolitan Police Department Web site.

While we wrote this book for those learning GIS from scratch, we also provide the material needed for becoming a crime mapping professional. We have used all our expertise in GIS education and crime mapping research to write a book that makes everything understandable, if not easy to learn. We have even included optional sections that use Microsoft Office Excel spreadsheet software for tasks that GIS packages cannot handle but are essential for becoming an expert crime mapper. We provide the results from these optional sections on the Maps and Data DVD, so you can skip these sections but still use the results in your work.

As a final note, this book uses real crime data obtained from the Pittsburgh Police Bureau and the Allegheny County 911 Center in Pennsylvania. Pittsburgh is in Allegheny County, so both sources of data are for the city of Pittsburgh. We protect confidentiality in this data by taking several steps: (1) we changed the year in dates, (2) we added random digits to house numbers in street addresses consistently across records, and (3) we replaced people's names with random names consistently throughout the records (using a random name generator by gender). We have worked with the Pittsburgh Police Bureau for more than 20 years and have built two crime mapping systems for them—one for narcotics detectives and the other for field officers (parts of which you use in chapter 3, build in chapter 4, and automate in chapter 10).

For the self-learner

If you are new to GIS, and crime mapping and analysis, we recommend that you work through the book from the beginning and go as far as you need to go. The book has three parts: part 1 on the basics of crime mapping, part 2 on selected crime analysis topics, and part 3 on building a crime mapping and analysis system from the ground up. You need to do all of part 1 first (chapters 1 through 4) on building and using crime maps and chapter 5 in part 2 on querying maps. Then you have the option of skipping chapters 6 and 7 in part 2 and jumping straight to part 3. You should do part 3 in chapter order (chapters 8 through 10), and then go back to chapters 6 and 7 in either order.

If you already know something about ArcGIS and crime mapping, you can easily jump to a chapter or chapters of interest. In fact, the book flow is in reverse in the sense that in parts 1 and 2, you use finished crime maps and systems, and then in part 3, you build the system you were using earlier. For that reason, we had to design a book that can be used "out of order."

Thus, all the necessary inputs that you'll need for tutorials and assignments in each chapter are available in computer folders on the Maps and Data DVD that comes with this book, even if you create them earlier or later than in the current chapter. Every chapter is self-contained, starting from the beginning of the subject at hand. While there is some repetition, in that a few steps from a previous chapter may be duplicated for the benefit of those doing a chapter out of sequence, most of it amounts to helpful reviews for those who did the earlier work.

For the instructor

You can use this textbook for a term course or semester course as well as for short courses. For a full semester course, there is time at the end for a student project. We recommend you obtain data from a police jurisdiction and have students build a crime mapping system. For a 10-week course, there is time for a short project. Additional work on crime analysis applications in chapters 5–7 provide good options for a short project.

Chapters 2–4 make a good module for those who want an introduction to GIS and crime mapping. Chapters 5–7 make a good module for students who are familiar with GIS but want more in-depth tools for crime analysis. Any of the chapters from 8 through 10 make good specialized modules or workshops for those needing behind-the-scenes, or "back office," skills—extracting and preparing basemaps and data from Web sources, geocoding address data, and building macros for process automation.

You can gain access to supplemental teaching materials, including a welcome letter for instructors, a sample syllabus, and solution files and grade sheets for assignments via an Instructor Resource DVD that is available on request at www.esri.com/esripress.

Summary of chapters

The chapters in *GIS Tutorial for Crime Analysis* follow a logical path for learning crime mapping and analysis, starting on how to use the system, and then on how to build it. That way, you get the big picture first, along with beginning GIS and crime mapping skills, and then as you progress through the book, you gain the knowledge to build the system and its components. In part 1, you start by learning how to use finished crime maps in ArcGIS, and then proceed to construct the same crime maps interactively by using prebuilt system components, including jurisdiction basemaps, police sectors, and master map files. In part 2, you apply your crime mapping skills to engage in sophisticated crime analysis. You learn how to carry out complex map queries, build crime map animations, and conduct hot spot analysis. Finally in part 3, you learn the back-office skills needed to produce analytical crime maps, including preparing and updating police jurisdiction basemaps, processing daily police report data into working map layers, and automating the entire process for efficiency and reliability.

If, however, you prefer to take a different path with the book—say, starting by conducting crime hot spot analysis in chapter 7, or even starting by using automation in chapter 10—this can work well, too. The book chapters are crafted with this scenario in mind, and so the output files for each tutorial are available on the Maps and Data DVD that comes with this book. If, for example, you were to start with chapter 10, the needed layer files for symbolizing pin maps for field officers, which are built in chapter 4, are available from the FinishedExercises folder for chapter 4; the needed master map layer of crime offenses, which is built in chapter 9, is available from the FinishedExercises folder for chapter 9; and so on. At each juncture of the book where you need files from an earlier chapter, you are instructed where to find them in the data files that come with this book if you did not perform the earlier work.

The following sections describe each chapter in greater detail.

Part 1: Working with crime maps

Chapter 1, "Introduction to crime mapping and analysis": This chapter defines crime mapping and analysis systems, including the basic skills needed as well as state-of-the-art and innovative procedures and methods that can improve your applications. In chapter 1, you will also find numerous references for additional reading.

Chapter 2, "Exploring ArcGIS Desktop": This chapter describes the crime mapping system you build and use in this book, and then provides exercises to get you started with the major components of ArcGIS Desktop—ArcMap and ArcCatalog. Catalog, which is available in ArcMap as a condensed version of ArcCatalog, is also explored. ArcMap is designed for building and using map documents, while ArcCatalog is a specialized utility program for creating, importing, and maintaining map files. In chapter 2, you gain experience using the interfaces of these components while exploring a crime map.

Chapter 3, "Using crime maps": This chapter builds on the overview of ArcGIS provided in earlier chapters to guide you on how to use the crime maps you will build in chapter 4 and automate in chapter 10. The maps you explore in this chapter address the needs of different end users, from field officers to specialized squads to top management and to the public. In chapter 3, you learn all the interactive skills needed to navigate and use ArcGIS map documents.

Chapter 4, "Designing and building crime maps": In this chapter, you build several map documents as templates that can be used to process various types of crime data. In chapter 4, you learn how to use cartographic principles to design effective pin maps and choropleth maps. You explore sophisticated designs for identifying patterns in recent crimes versus older crimes, the use of size-graduated point markers to indicate the number of crimes at a location, custom numerical scales that are most suitable for crime maps, visible scale ranges for drilling down to see crime details, and the randomization of crime point locations to protect personal privacy in maps that are made available to the public.

Part 2: Conducting crime analysis

Chapter 5, "Querying crime maps": This chapter covers the fundamental approaches to crime analysis through attribute and spatial queries. In chapter 5, you learn how to use the ArcMap Query Builder interface to generate the major kinds of attribute queries crime analysts use—date range, crime type, day of the week, time of day, and person/object characteristics. Then you learn how to use spatial queries and buffers to select crimes that are near or proximate to other features for proximity analysis to discover crime-prone land uses.

Chapter 6, "Building crime map animations": This chapter covers how to build map animations to study the dynamics of serial crimes and the mechanics of crime hot spots. ArcMap automates the building of individual frames, or map images, that can be played one after another for a "map movie." Chapter 6 includes two key principles we've developed in our research for building effective crime map animations: (1) retaining selected recent crimes to serve as context for newer crimes and (2) annotating important crime patterns as they first emerge.

Chapter 7, "Conducting hot spot analysis": This chapter covers widely used approaches for resource allocation, including the leading hot spot estimation method, kernel density smoothing, which is available through the ArcGIS Spatial Analyst extension, and use of the Spatial Statistics toolbox available in ArcToolbox in ArcMap. The kernel density smoothing method you learn in this chapter is one of the best available for identifying and displaying crime hot spots. Chapter 7 also covers the nearest-neighbor test for spatial clustering and how to use the Getis-Ord Gi* test for hot spot analysis.

Part 3: Building a crime mapping and analysis system

Chapter 8, "Assembling jurisdiction maps": This chapter first teaches how to identify and download basemaps and census data from Internet sources, and then goes through the steps to modify and import these materials into ArcGIS. In chapter 8, you learn how to extract and build additional maps from the basemaps to create police jurisdiction basemaps. You also learn how to create police sector maps by "dissolving" census tract boundaries to combine the tracts or by using on-screen streets as a guide to digitize desired areas. Finally, you learn how to digitize block watch areas with the use of a tracing tool.

Chapter 9, "Preparing incident data for mapping": This chapter shows how to use ArcMap geocoding to process police report data into maps. Geocoding matches the street addresses in police reports to the addresses on a digital street map, thereby assigning incidents to points on a map. The street address data that goes into police reports often has inaccuracies or inconsistencies, so the ArcGIS application for interpreting and standardizing address data fixes many of these problems. Chapter 9 teaches advanced skills for improving the accuracy of geocoding, such as the use of external sources for addresses and the repair of faulty street maps. Because crime mapping also makes use of aggregate data, such as monthly crime counts by police sector, this chapter covers spatial joins and aggregation to create such data. Lastly, chapter 9 teaches you how to append newly geocoded point and aggregate data to master map layers. Master layers are used for the inputs for all crime mapping and analysis applications.

Chapter 10, "Automating crime maps": This chapter allows you to take your GIS skills to a new level through the use of the ArcToolbox and ModelBuilder packages for creating macros to build new programs. These technologies automate most of the steps for processing police data into GIS components. They also automate production of the crime maps you build in chapter 4 and the master crime map layers you create in chapter 9. In chapter 10, you learn how to use ModelBuilder to create macros, or models, for building crime maps through a visual flowchart interface; how to fix "bugs," or errors; how to document macros for end-user help; and how to configure the user interface so that the inputs can be changed when running a model. The result is that you can build an automated crime mapping system for a police department entirely with off-the-shelf ModelBuilder, and without the need for custom programming.

Prerequisites for using this book

We have taught GIS to a wide range of students, all the way from underprivileged high school students in a successful after-school and summer school program (see the InfoLink Web site for Heinz College, Carnegie Mellon University) to undergraduate and graduate students and GIS professionals—all at Carnegie Mellon University in Pittsburgh. These student bodies have always had motivated students from a great variety of backgrounds, so our approach is to start at the beginning to get everyone on board, and then move forward to the point where students can become self-sufficient in using ArcGIS. Regardless of your background in GIS, we believe you can be successful with the use of this book.

Of course, you'll need a certain level of computer literacy to work effectively with GIS—a knowledge of how to work with files and folders on PCs as well as how to use Microsoft Office, especially the Word and PowerPoint packages, because these are often the media for map output from GIS. Some data preparation tasks for crime mapping are best done using Microsoft Office Excel, so the book includes step-by-step tutorials for this spreadsheet package. If you wish to skip the work in Excel, you can use the finished files contained on the Maps and Data DVD that comes with this book to proceed with the next steps.

Microsoft Windows software has some capacity for compressing (or zipping) files as well as extracting files from compressed files, which will be useful in working with this book. Search "zip" in Windows Help for more information. Nonetheless, we recommend that you buy a utility program for this purpose. If you are in a classroom setting, you need to learn how to compress and preserve folders and subfolders, with all their contents, in a single file for use in submitting assignments. It is also beneficial to learn how to extract original files from compressed files when downloading basemaps and data, which you do in chapter 8.

To do the exercises and assignments in this book, you need to have ArcGIS Desktop 10 software installed on your computer, or else download a trial version of the software. See appendix F for instructions on how to download the software. You will need the code printed on the inside back cover of this book to access the download site.

You must also install the Maps and Data DVD that comes with this book on your computer. See appendix E for the data license agreement and appendix F for instructions on how to install the data. The Maps and Data DVD contains all the starting map documents, spatial data files, and map layers that are needed to complete the tutorial exercises and assignments, although it does not include the software itself. The Maps and Data DVD contains the folders for you to store your work, including exercises and assignments. It is important that you follow the instructions in appendix C on where to store your files, especially if you are in a class. Then, when you turn in your assignment files and folders, the relative paths to stored resource files (map layers) will be correct, so whoever grades your work can open your files and get your solutions without getting a copy of your basemap layers.

Acknowledgments

We are most grateful to the many people who helped us with this book. First, we wish to thank Lew Nelson, Industry Solutions department manager and law enforcement industry manager at Esri, for encouraging us to write this book. He wanted a book that not only covers the basics of crime mapping and analysis, but that also has the potential to advance the state of the art in the field. We endeavored to take on Lew's challenge and hope that we have succeeded in some measure. Secondly, we are grateful as always to the editing and production team at Esri Press.

Many thanks go to Chief Nathan Harper of the Pittsburgh Police Bureau and Chief Robert A. Full of the Allegheny County Department of Emergency Services in Pennsylvania for providing the data used in this book. Detective Deborah Gilkey, of the Crime Analysis/Intelligence unit of the Pittsburgh Police Bureau, participated in some of our research that informed the writing of this book as well as provided us with some important data collections that are used in the book.

Learning pathways

GIS Tutorial for Crime Analysis is designed for use by self-learners as well as by students in the classroom. By using this book on your own and doing the tutorial exercises and assignments, you should be able to learn crime mapping and analysis skills, and then be able to apply them in your work within a police department. This section explains the path that will help you succeed in using this book to learn crime mapping and analysis skills.

Prerequisites

To be successful with *GIS Tutorial for Crime Analysis*, you need to be able to work with files and folders on a computer, plus have a working knowledge of Microsoft Office Word and Microsoft Office PowerPoint applications. We recommend that you also have some experience with Microsoft Office Excel applications, but this is not essential. The tutorials contain complete step-by-step instructions on use of the spreadsheet software.

Getting started

Install the downloadable 180-day trial version of ArcGIS Desktop 10 software on your computer by using the code inside the back cover of the book, if you do not have other access to this package. Instructions on downloading the software are in appendix F.

Then install the Maps and Data DVD from the back of the book on your computer. This DVD has the starting map documents for each chapter, plus all map and other data files you need to complete the tutorials and assignments. It would be best to install the data on your C drive if you have room, but if not, another drive will work fine. Instructions on downloading the data are also in appendix F. Also, please see appendix E for the data license agreement.

If you are working in a computer lab on public computers, it is best to save all your GIS files on a thumb drive and work from that.

What to do if you already have GIS and crime mapping experience

If you know some parts of ArcGIS and crime mapping, you can easily jump to a chapter or chapters that interest you. In fact, the flow of the book is in reverse, in the sense that you use finished crime maps and systems in parts 1 and 2 of the three-part book. Then in part 3, you build the system you were using earlier from scratch. So, we had to design a book to be used "out of order." Thus, every chapter has all the necessary inputs to tutorials and assignments in computer folders on the Maps and Data DVD, even if the book has you create them earlier or later than the current chapter. Every chapter is self-contained.

Every chapter starts from the beginning of the subject at hand, and there is some repetition of steps from earlier chapters for the student jumping to a chapter. The limited repetition amounts to helpful review for those who did the earlier work while it benefits those who didn't.

Part 3 should be especially attractive to the experienced GIS user. There, you learn all the "back room" workflows and skills to build a crime mapping and analysis system. We have built crime mapping and analysis systems for several police departments, and we have included much of that throughout the book, especially in part 3.

What to do if you are new to GIS and crime mapping

If you are new to GIS, and crime mapping and analysis, your best bet is to work through the book from the beginning and go as far as you need to go. The start of your journey in part 1 is the easiest way to get into the field. You learn how to use prebuilt crime maps, and then in chapter 3, you actually build the same maps from prebuilt map layers. In part 3, you build those layers from scratch.

Part 2 opens new horizons for those who have gained basic crime mapping skills from part 1. In part 2, you learn GIS skills and workflows to conduct advanced crime analysis, starting with a comprehensive approach to querying mapped crime incidents. Next is a topic you will not find in other GIS textbooks—how to build map animations. The animations are for crime analysis of serial crimes and for hot spot analysis—the two major kinds of analysis that police departments need. Animations are easy to build in ArcGIS Desktop 10 and provide a new tool for crime analysis that offers a competitive advantage. Finally, part 2 includes a chapter on hot spot analysis using a leading technology, kernel density smoothing.

While it is best to have part 1 done before going on to part 3, you can skip part 2 if you like and return to it later. In part 3, you learn how to build every aspect of a crime mapping system, starting with downloading free basemaps and processing them into the critical jurisdiction basemaps a police department needs. Then you learn the important subject of geocoding crime incident street addresses into points on a street map, with all the "tricks of the trade" for getting the best results. Finally, something you may not find in any other GIS textbook is the use of the ArcGIS macro application, ModelBuilder, to build an automated system, which of course in this case is a crime mapping and analysis system. In ModelBuilder, you use a simple drag-and-drop interface and computer forms that you fill out to build a flowchart model of the processes needed for crime mapping.

Selected readings

You can learn much about crime mapping and analysis approaches and theories by working through our book. After you finish a chapter, it is a great time to delve into the related literature once you have seen the material implemented with real data. A few suggested sources from our own library shelf follow. Readings marked with an asterisk (*) are free downloads and are especially helpful.

Crime mapping and analysis

*Anselin, L., J. Cohen, D. Cook, W. L. Gorr, and G. Tita. 2000. "Spatial analyses of crime," in *Measurement and Analysis of Crime and Justice*, ed. D. Duffee. *Criminal Justice* 4: 213–62. Washington, D.C.: National Institute of Justice.

Boba, R. L. 2005. *Crime analysis and crime mapping*. Thousand Oaks, CA: Sage Publications.

Bowers, K. 2007. *Mapping and analyzing crime data*. London: Taylor and Francis.

Chainey, S., and L. Thomson. 2008. *Crime mapping case studies: Practice and research*. West Sussex, UK: John Wiley and Sons.

Chainey, S., and J. Ratcliffe. 2005. *GIS and crime mapping*. West Sussex, UK: John Wiley and Sons.

*Eck, J. E., S. Chainey, J. G. Cameron, M. Leitner, and R. E. Wilson. 2005. "Mapping crime: Understanding hot spots." NIJ Special Report. Available from `http://www.ncjrs.gov/pdffiles1/nij/209393.pdf`.

*Harries, K. 1999. "Mapping crime: Principle and practice." NIJ Special Report. Available from `http://www.ncjrs.gov/pdffiles1/nij/178919.pdf`. (Note that the crime map on the cover, in the center and that has grid cells, was created by one of the authors of this textbook.)

Paulsen, D. J., and M. B. Robinson. 2008. *Spatial aspects of crime: Theory and practice*. 2nd ed. Boston: Pearson.

Weisburd, D., and T. McEwen. 1998. *Crime mapping and crime prevention*. Monsey, NY: Criminal Justice Press.

Cartographic and graphic design

"Color meaning." In *Color wheel pro*. `http://www.color-wheel-pro.com/color-meaning.html` (accessed December 7, 2010).

Brewer, C. A. 2005. *Designing better maps: A guide for GIS users*. Redlands, CA: Esri Press.

"Color principles: Hue, saturation, and value." Graphic Communications Program, North Carolina State University, College of Education. `http://www.ncsu.edu/scivis/lessons/colormodels/color_models2.html#saturation` (accessed December 7, 2010).

MacEachren, A. M. 1994. *Some truth with maps: A primer on symbolization and design*. Washington, D.C.: Association of American Geographers.

Tufte, E. R. 2001. *The visual display of quantitative information*. 2nd ed. Cheshire, CT: Graphics Press.

*Free downloads.

Part 1
Working with crime maps

OBJECTIVES

Learn about crime mapping and analysis systems

Chapter 1

Introduction to crime mapping and analysis

The objective of this computer workbook is to teach you how to build and apply a crime mapping and analysis system with the use of ArcGIS Desktop 10 software. This chapter explains how to use this book, whether you are a self-learning professional or a student in a class. It describes the basics of crime mapping and analysis and introduces the state-of-the-art and innovative features detailed in this book. *GIS Tutorial for Crime Analysis* is geared toward learning by doing, but we also recommend additional readings on the environmental criminology theories that underlie the ArcGIS applications used in this book.

Learning the system

Mapping and analyzing crimes requires digital crime data, digital maps, software, computers, police personnel, procedures, methods, and processes all working together—in other words, it requires a fully functioning geographic information system (GIS). This book makes you a user and then a builder of crime mapping and analysis systems that can efficiently and effectively further the day-to-day work of a police department and broaden its long-range scope of citizen protection and crime prevention.

Crime mapping and analysis systems

Applying a GIS to crime mapping and analysis is a major innovation of police organizations around the world. It enables crime analysts, uniformed officers, investigators, intelligence officers, and police executives to access and analyze data in crime maps so they can better enforce the law and prevent crimes.

This integration of multiple data sources into crime mapping and analysis systems is just part of the good news. Another is that this book provides a path, from start to finish, for harnessing GIS technology and ArcGIS Desktop for use in crime mapping and analysis applications.

Unlike the many GIS textbooks that deal with planning and special projects, this book is aimed at the operations and tactical levels of service delivery, part of the inner fabric of police departments. Every approach and every method used in this book is aimed at building a crime mapping and analysis *system* that will benefit the many needs of a police department. This system is designed to handle the constant flow of new crime data as input, to process all the crime data into spatial formats, and to ultimately produce crime maps and other spatial outputs for different groups of users. The "back office" subsystem handles the inflow of new crime data from police reports, geocodes it to give it location coordinates, and appends it to the original master crime map layer to produce the updated or current master file. Jurisdiction basemap layers that include streets are needed to geocode addresses into points on the map. Jurisdiction basemaps and the current master crime map layer serve as inputs for all manner of crime mapping and analysis functions, from analytical mapping to map animations. ArcGIS makes it possible to build this system, and this book makes it possible for you to learn and use ArcGIS so you can build it.

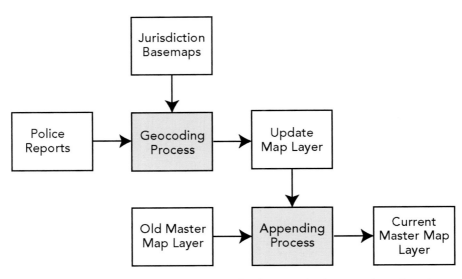

The workflow for geocoding updated crime data and appending it to create the current master crime map layer.

The production of crime maps and analyses for different groups of users.

This book's approach is to first have you use a crime mapping and analysis system, and then learn how to build it. That way, you get the big picture first, along with the motivation to acquire the deeper knowledge and skills needed to build the components that make up the system. This sequence is also in line with typical job expectations in which you first use GIS methodology, then learn more sophisticated applications, and finally, build and update the underlying system.

The next section of this chapter provides definitions of crime analysis, GIS, and crime mapping. Following sections describe the basics of crime mapping, as well as state-of-the-art criminology concepts and tools, and where you can find these subjects covered in the book. The last section suggests references for the theories of environmental criminology that guide crime mapping and analysis.

Definitions used

Working definitions for the organizational activities covered in this book and for the major tools used can help your study of GIS in crime mapping and analysis. The definitions, analogous to an organization's mission statement, can help guide you in the day-to-day operational tasks of a police department, as well as help define your overall purpose in using GIS for crime mapping and analysis. So, for example, when using GIS for crime mapping and analysis, you do not just produce maps, you use them to help enforce laws and prevent crimes. R. L. Boba (2005, 6) provides a good start with an overarching definition of crime analysis:

> *Crime analysis* is the systematic study of crime and disorder problems as well as other police-related issues—including sociodemographic, spatial, and temporal factors—to assist the police in criminal apprehension, crime and disorder reduction, crime prevention, and evaluation.

This definition is helpful because it states the purpose of crime analysis (criminal apprehension and prevention), as well as how it is accomplished—through the systematic study of crime, including sociodemographic, spatial, and temporal factors.

We define a *GIS* as a computer application that stores, retrieves, and displays spatial data on maps. Key to GIS is the set of spherical world coordinates, latitude and longitude, that can locate any spatial feature in the world on the surface of the globe. Spatial features, along with their related data records, can also be represented by graphics or images. So, for example, a crime location can be represented as a point on the map that has geographic coordinates as well as an event record that contains the date and time of occurrence, the street address of occurrence, and the crime type. To display these coordinates on a computer screen or paper map, latitude and longitude coordinates must be transformed, or projected, to flat coordinates. A GIS handles such projections in either direction, from spherical to flat, and vice versa. For more information on map projections, see Caitlin Dempsey's definition of map projections, available on the GIS Lounge Web site.

We define *crime mapping* as an application of GIS to assign spatial coordinates to crime incidents and other locations such as criminals' residences and to produce map compositions with crime locations and spatial context features such as streets and police sectors. Crime mapping produces planned, periodic maps for various audiences, such as field officers, investigators in various crime squads, and the public. For example, crime mapping produces daily pin maps that pinpoint crime locations for field officers and monthly maps that are used in CompStat meetings to pinpoint crime trends. CompStat, short for "computer statistics," is a process, initiated by the New York City Police Department and now used around the world, to review police operations and allocate resources. Crime mapping also produces situational crime maps that are used to assess unique problems such as serial robberies and gang rivalries. Crime mapping is a required input for the many crime analysis tools and analyses included in this book.

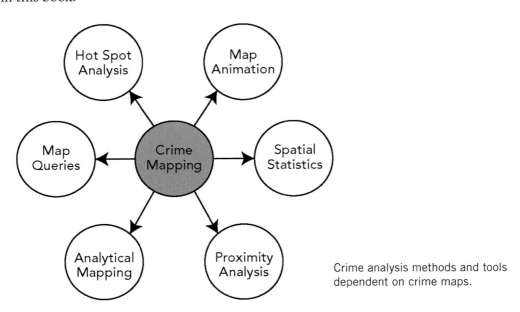

Crime analysis methods and tools dependent on crime maps.

Mapping and analyzing crime

This book provides crime mapping and analysis knowledge and skills at two levels. First, it covers the basics of GIS and how GIS relates to crime mapping and analysis. Then it pushes the envelope and introduces state-of-the-art, innovative material from the field. This section provides overviews of both the basics of crime mapping and analysis, as well as the innovations that are introduced in this book.

Crime mapping basics

This book covers the basics of crime mapping to give you a good foundation in the field. You will need to do some external reading (see the references provided later in this chapter and throughout the book) to gain in-depth background on relevant criminology theories, but many of these theories are introduced here. This section describes the basic knowledge and skills you will gain as a result of this book, as well as the rationale behind this methodology. The basics of crime mapping are scattered throughout the book, presented in step-by-step exercises in a logical instructional sequence, so this section also tells you where to look in the book for instructions on particular skills and methodologies.

Also, as needed, you can refer to appendix A, "Tools and buttons," and appendix B, "Task index," to find particular crime mapping and GIS tools and tasks.

Systems analysis and design: Most critical is that you learn how to identify all the potential users of crime maps and analysis, determine their needs by interviewing users and reading the literature, and then be able to provide maps and analysis that meet their needs. This work includes making prototype maps, getting feedback, and obtaining approval from management for the creation of official crime maps. We followed this process ourselves in building several crime mapping and analysis systems for police departments, always being careful to interview chiefs of police down through the ranks to field officers to determine their needs. Chapter 3, "Using crime maps," makes use of the best crime maps that we and our police collaborators built, illustrating how to meet the needs of three types of audiences: field officers, investigators and top managers, and the public. You'll need this kind of systems analysis and design work on a continual basis to meet additional crime analysis needs as they arise and to add new audiences, such as special task forces and the media, for the crime maps you build.

Cartographic design: The crime maps you design and produce as a result of this book have an analytical purpose—to identify crime patterns that are targets for law enforcement and crime prevention. While crime maps should be attractive in their design, their primary purpose is to convey information effectively. Thus, to be a good map designer, you need to learn certain cartographic (map design) principles. Most critical is the principle of graphic hierarchy—that is, drawing attention to crimes, for example, with the use of color, while placing second-ary, spatial-context features such as streets in gray or drab "background" colors. Chapter 4, "Designing and building crime maps," illustrates this principle in the maps you build in those exercises. In chapter 4, you also learn how to become a sophisticated user of special crime point markers, which are part of ArcGIS symbology for crime mapping and analysis. In par-ticular, you are asked to modify these point markers and save them for future use in layer files. You also learn how to use monochromatic color ramps, for shading, to color-code area maps (called "choropleth" maps) to convey numeric scale for quantitative attributes such as crime level or population. Such color ramps clearly convey the order of magnitude through color value, or the amount of black in a color, while retaining that meaning when maps are converted into a black-and-white photocopy. Finally, you learn how to use dichromatic color ramps to color-code choropleth maps for changes in crime levels, such as shades of blue for decreases in criminal activ-ity and shades of red for increases in criminal activity. Blue and red are far apart on the color wheel (more on color selection in chapter 4), so using these contrasting colors can vividly portray crime decreases versus increases on a choropleth map.

Attribute and spatial queries: A GIS stores data to meet many organizational needs. A major part of your responsibility, as a crime mapper and analyst, is to extract information from data as part of that system. Much of that work is accomplished through queries that extract selected data records and attributes from master crime map layers. Chapter 5, "Querying crime maps," teaches the logical criteria or filtering part of the Structured Query Language (SQL), the uni-versal language for attribute queries, so you can select these records and extract the needed information from them. The method of instruction is itself structured by focusing on the types of questions or queries that you need to make, starting with "what," "when," and "where" and following with "who" and "how." Moreover, you learn how to save your queries as files that can be reused or modified as new crime mapping and analysis needs arise. Unique to GIS is a second way to make queries, beyond SQL queries. This second technique is based on the loca-tions, or coordinates, of mapped features that are in proximity to other features, and thus can serve as a means of revealing important crime patterns and trends. It is critical in crime analysis to learn how to create buffers around locations such as bars, ATMs, schools, and gas stations so you can retrieve information on crimes that occur near these features. For example, the buffer of a point is a circle around it, and you prescribe the radius. Chapter 5 covers this material so you can keep track of sites that attract crime by placing buffers around them, and then analyze those buffers.

Jurisdiction maps: If you are introducing crime mapping to a police department or want to update jurisdiction basemaps or add new basemap layers to an existing crime mapping system, you need to know how to extract or combine basemaps. You can download these basemaps for free from the Internet or get them from vendors who augment government-provided basemaps or create their own specialized layers. Chapter 8, "Assembling jurisdiction maps," covers downloading basemaps as well as spatial data, including street centerlines, political boundaries, water features, census tracts, and census data. Generally, a police jurisdiction does not unilaterally match the geographic boundaries of downloaded basemaps, but is composed of subsets or collections of basemaps. Thus, it is essential to learn how to extract and assemble jurisdiction basemaps to meet a police department's needs. Police departments also have unique boundaries of their own that you may need to create or update, including police sectors (car beats), community policing areas, and block watch areas. Aerial photographs can be valuable for providing detailed spatial context, and thus be a good background for crime maps, but they have enormous file sizes. Instead of using these outsize layers from your own computer, you can add them to your maps as Web services provided over the Internet via "cloud computing." In chapter 2, "Exploring ArcGIS Desktop," you learn how to use maps from ArcGIS Online to provide this spatial context.

Geocoding: Every minute and every hour of the day produces new computer-aided dispatch (CAD) calls for service, new field contact reports, new offense reports, and new arrest reports that a police department must process. This flow of tabular data needs to be spatially enabled (geocoded) by assigning spatial coordinates to crime locations, which is covered in chapter 9, "Preparing incident data for mapping." Then this data can be mapped as crime points. ArcGIS has very sophisticated geocoding (or address matching) algorithms that assign street coordinates to street address data. These coordinates are derived from street centerline basemaps. A second phase of geocoding aggregates this point data to space and time series data—for example, it can add up the burglaries per week for each police sector. GIS processing includes a unique GIS function, called a spatial join or overlay, that can assign a police sector to all resident burglary locations, which you learn to use in chapter 9. You are then in position to count up all burglaries per week per police sector for use in a new file. This space and time series data is essential for studying crime trends and producing the type of choropleth maps that are used for the early-warning crime system you develop in chapter 4.

State-of-the-art concepts and tools

While this book provides a solid introduction to crime mapping and GIS, it also highlights innovative crime analysis concepts and state-of-the-art technology. Some of the major elements of this book include creating and updating master crime files, using macros to automate processes, animating crime maps, using and fine-tuning hot spot analysis, and furthering the goals of predictive policing to fight crime more effectively.

Master crime map layers: The data-processing world has been creating master files for transactional data, such as sales data, for more than 40 years. A master file holds historic data that is kept up-to-date by the continuous appending of update files. The master file is a valuable asset that provides a single place to get transactional data for use as inputs to a variety of applications. This same kind of data organization is needed for crime data to get the type of common, file-based crime mapping and analysis systems developed in this book. Yet instead, many current systems have ad hoc collections of files that follow the idiosyncratic patterns of individual crime mappers. In chapter 9, you learn how to use best practices to build and update master crime map layers to keep the crime mapping and analysis system organized and usable.

Macros for crime map automation: Chapter 10, "Automating crime maps," gives you the ability to harness macros in ArcGIS using ModelBuilder to automate the processes used in the production of periodic crime maps—from geocoding crime reports to updating master files to outputting crime maps. ModelBuilder technology allows you to use computer macros to build models, or programs, to automate processes. In chapter 10, you learn how to use the ModelBuilder interface to build a flowchart by dragging in tools, each with its own inputs and outputs. Each tool has a computer form in which you can set properties or parameters to adjust the model for new inputs and outputs. Chapter 10 is the first text to put ModelBuilder to work in a crime mapping system.

Crime map animations: Animating crime maps helps put crimes in context by showing a spatial-temporal sequence of where and when crimes occur. In chapter 4, you learn to include a time context in crime maps, thus distinguishing newer crimes from older crimes. That way, you can see some of the dynamics of crime patterns. Chapter 6, "Building crime map animations," takes mapping the dynamics of crimes to a new level, offering the first instructional material available on how to do crime map animations. Although it is fairly simple to construct animations in ArcGIS, this book shows how to use crime map animations in the study and analysis of serial crimes and crime hot spots for more effective portrayal of these patterns.

Gold-standard-based hot spot analysis: Ever since the first computerized crime maps were produced, crime analysts and researchers have been struck by how concentrated crime is in urban areas. A small percentage of the neighborhood blocks in a jurisdiction can account for 30% to 50% of the crime. Spatial crime clusters are known as hot spots, and they have become favored targets for law enforcement and crime prevention. One of the goals of hot spot analysis is to use an objective method for identifying current hot spots, and arguably, the best method for identifying hot spots is kernel density smoothing. Kernel density smoothing uses crime points as inputs to generate a 3D crime density surface, where the resulting peaks are the crime hot spots. One problem in calibrating kernel density smoothing to choose the right area for hot spots is choosing the value for the search radius. Until the material discussed in chapter 7, "Conducting hot spot analysis," was developed, there was no good way of choosing this value. Chapter 7 gives you a good place to start by using visual inspection to create a sample of "gold standard" hot spots, manually drawing boundaries around these areas. You can then train or calibrate kernel density smoothing to find hot spots by trying different values for the search radius. The radius that approximates the peaks, matching the gold standard hot spots that you have predefined, is thus chosen as best. So, in practice, calibrated kernel density smoothing replicates "expert judgment," allowing you to find hot spots at the push of a button, which saves time and produces consistent results.

Predictive policing: "Predictive policing" is a new model for policing that uses crime pattern changes and forecasts to aid in crime prevention. One of the authors of this book is a pioneer in crime forecasting and has conducted research on attempting to predict where the next hot spot will emerge or to detect it as soon as it appears to allow police departments to nip crime in the bud. Innovative map designs are scattered throughout the book that deal with the dynamics of hot spots, including an early-warning system of crimes as well as leading-indicator crime maps, which you build in chapter 4.

Environmental criminology references

Whether you are a self-learning professional or a student enrolled in a course, you will benefit from studying the environmental criminology literature on crime mapping and analysis. There is no correct sequence in what to learn first, whether theory or hands-on skills. If you are a student in a class, your instructor will weave both theory and skills into the class. If you are a self-learner, you can study both, or first learn one and then the other. You can expect to comfortably get through this workbook and learn some theory by doing, and then follow up through self-directed reading from the literature.

Underlying the success of crime mapping are several insightful environmental criminology theories on how place affects crime (and vice versa), including broken-windows theory, routine-activity theory, travel to crime, and crime displacement. These theories are covered in a variety of books and other resources, so this workbook does not repeat them here. Instead, it is recommended that you purchase one or more supplemental theoretical books such as *Crime Analysis and Crime Mapping* (Boba 2005); *Mapping and Analyzing Crime Data* (Bowers 2007); *Crime Mapping Case Studies: Practice and Research* (Chainey and Thomson 2008); *GIS and Crime Mapping* (Chainey and Ratcliffe 2005); *Spatial Aspects of Crime: Theory and Practice* (Paulsen and Robinson 2008); *Geographic Profiling* (Rossmo 2000); and *Crime Mapping and Crime Prevention* (Weisburd and McEwen 1998).

Outstanding materials are also available free from key Web sites. The earliest and most influential national organization on crime mapping is the Mapping and Analysis Program for Public Safety (MAPS) under the National Institute of Justice in the U.S. Department of Justice, Office of Justice Programs. The program is the brainchild of Dr. Nancy LaVigne. On the National Institute of Justice Web site under Publications, you can find many resources for downloading, including "Mapping Crime: Principle and Practice" (Harries 1999); "Mapping Crime: Understanding Hot Spots" (Eck et al. 2005); and "Measurement and Analysis of Crime and Justice: Spatial Analyses of Crime" (Anselin et al. 2000). This Web site also offers access to CrimeStat, free statistical software sponsored by MAPS and developed by Ned Levine and Associates. CrimeStat uses Esri formatted map files as inputs and outputs for specialized statistical methods used in crime analysis. For an additional source of free crime analysis materials, go to the Jill Dando Institute of Crime Science page on the University College London Web site and click the Publications link.

References

Anselin, L., J. Cohen, D. Cook, W. L. Gorr, and G. Tita. 2000. "Spatial analyses of crime," in *Measurement and Analysis of Crime and Justice*, ed. D. Duffee. *Criminal Justice* 4: 213—62. Washington, D.C.: National Institute of Justice.

Boba, R. L. 2005. *Crime analysis and crime mapping.* Thousand Oaks: Sage Publications.

Bowers, K. 2007. *Mapping and analyzing crime data.* London: Taylor and Francis.

Chainey, S., and J. Ratcliffe. 2005. *GIS and crime mapping.* West Sussex, UK: John Wiley and Sons.

Chainey, S., and L. Thomson. 2008. *Crime mapping case studies: Practice and research.* West Sussex, UK: John Wiley and Sons.

Eck, J. E., S. Chainey, J. G. Cameron, M. Leitner, and R. E. Wilson. 2005. "Mapping crime: Understanding hot spots," *NIJ Special Report.* Available at `http://www.ncjrs.gov/pdffiles1/nij/209393.pdf`.

Harries, K. 1999. "Mapping crime: Principle and practice," *NIJ Special Report.* Available at `http://www.ncjrs.gov/pdffiles1/nij/178919.pdf`.

Paulsen, D. J., and M. B. Robinson. 2008. *Spatial aspects of crime: Theory and practice.* 2nd ed. Boston: Pearson.

Rossmo, D. K. 2000. *Geographic profiling.* Boca Raton, FL: CRC Press.

Weisburd, D., and T. McEwen. 1998. *Crime mapping and crime prevention.* Monsey, NY: Criminal Justice Press.

OBJECTIVES

Explore the ArcMap user interface
Explore the ArcCatalog user interface
Examine map layer properties
Examine Layout View

Chapter 2

Exploring ArcGIS Desktop

ArcGIS is a collection of software packages that comprises a vast and exhaustive storehouse of GIS functions for building and using electronic maps of any kind. Consequently, it has a large and complex user interface. This chapter provides a hands-on introduction to the user interfaces for ArcMap and ArcCatalog, the major components of ArcGIS Desktop, as you explore an informative crime map.

Overview of ArcGIS Desktop software

ArcGIS Desktop is one of the world's largest Windows-based application programs. It has an intricate user interface that you cannot simply open and expect to figure it out intuitively. This is one of the many reasons you need this book. The step-by-step exercises in this chapter start off by having you become familiar with the user interfaces for ArcMap and ArcCatalog, the primary components of ArcGIS Desktop. ArcMap is for building, editing, and displaying map documents, while ArcCatalog is a specialized utility program for creating, importing, and maintaining map files.

You need to have ArcGIS Desktop and this book's data installed on your computer to carry out the steps in the exercises throughout this book. See appendix F for instructions on downloading a trial version of the ArcGIS Desktop 10 software as well as installing the Maps and Data DVD that comes with this book. Note that you can install the trial software package only once. Before you open this program on your computer, it is helpful to have some background knowledge, which is what comes next.

Map documents

The ArcMap component of ArcGIS Desktop uses a map document approach to GIS. In this book's exercises, you build and use map documents in ArcMap. A map document is a file (with the file extension ".mxd") that stores a map composition consisting of several map layers. The ArcMap user interface has several features you will explore in these exercises.

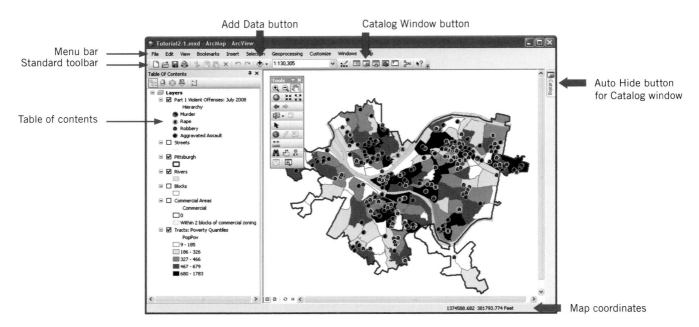

The ArcMap interface includes a map and a table of contents that lists the map's layers.

Each map layer can have separately stored symbolization. For example, a classic crime "pin map" depicting crime points, shown in the figure, may have the following three map layers: (1) crime points for a week, (2) street centerlines, and (3) police sector boundaries. As map designer, you could add more layers if you wish; for example, you could add crime-prone locations such as bars, liquor stores, and convenience stores. (These additional layers might help explain crime patterns and lead to better crime prevention.) All map layers overlay each other correctly on the map because map layers have coordinates linked to the unique world coordinates of latitude and longitude. Notice that map layers each have homogeneous features—that is, all crimes, all streets, all police sectors, or the like.

Symbolization refers to the shapes, sizes, and colors of graphic features on a map that are chosen to help convey information about crime patterns. The symbolization in the figure consists of: (1) size 10 point markers, with specialized crime analysis point markers and selected bright colors for crimes; (2) size 1, 50% gray lines for streets with street names used as labels; and (3) size 2, black boundaries for police sectors with police sector numbers used as labels. Once you symbolize a layer such as crime points by using the graphic user interface in ArcMap, you can save the symbolization to a file and reuse it to show next month's crimes without having to go through all the interactive steps of symbolizing again. This capacity allows you to build map templates—symbolized map documents that can be reused with minimal work. In fact, as you learn in chapter 10, you can reuse map templates to automate map production, building the map for next month "at the push of a button" by using the macro functionality in ModelBuilder, an application in ArcMap.

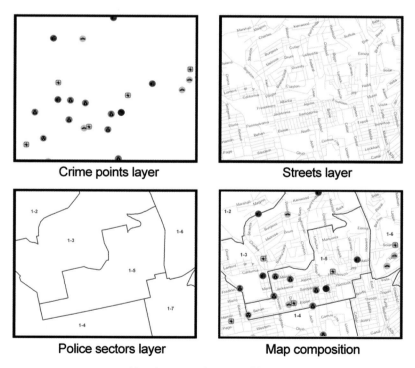

Crime points layer   Streets layer

Police sectors layer   Map composition

Map layers and composition

Learning about ArcGIS

ArcMap is the major GIS application package in the ArcGIS Desktop suite. It has the functionality to create a map document, which can be displayed or output as a file for a printed document, can be navigated to explore an area, and can be queried for information and analysis.

Tutorial 2-1

Exploring the ArcMap user interface

To get started, you will open a completed map document with crime and other data to get the feel for ArcGIS and crime mapping.

Open map document

1   From the Windows Start menu, click All Programs > ArcGIS > ArcMap 10.

2   In the ArcMap window, click Existing Maps > Browse for more.

3   Go to the drive where you installed the EsriPress\GISTCrime\Maps folder (for example, C:\EsriPress \GISTCrime\Maps), select Tutorial2-1.mxd, and click Open. The resulting map of Pittsburgh, Pennsylvania, displays population and crime data via a choropleth map of census tracts (a map consisting of areas coded by color) and a map of crime points. The choropleth map uses a monochromatic gray color ramp for the population below the poverty line. A color ramp has continuously changing colors or shades of darkness (called "values") for use in identifying categories of numerical order. The point map uses point markers coded by crime type to depict Uniform Crime Report (UCR) Part 1 violent crimes (murder, rape, robbery, and aggravated assault) for July 2008. A UCR crime is the highest offense in the FBI listing hierarchy for a crime incident, although more than one type of offense may have been committed in the same incident. An incident that included both a robbery and an aggravated assault, for example, would be recorded as a robbery, because robbery is the higher-level, or more serious, crime. It is well known that serious violent crime tends to occur in high-poverty areas—a trend that is evident in this map.

4   Click File > Save As. Go to your chapter 2 folder in MyExercises and click Save. You will regularly save your working map document to a chapter folder such as chapter 2 in MyExercises. ArcGIS automatically stores any files you create to the same working folder so you will have these files when and where you need them. This practice also preserves the starting map documents provided in the Maps folder in original condition, so if you wish to repeat a tutorial, you have the correct starting map document. Lastly, chapter folders in the FinishedExercises folder of MyExercises have finished map documents and other files created in a tutorial so you can view or use these materials as needed. Later exercises often use files created in earlier exercises, and you will always be able to find them in the FinishedExercises chapter folders if you did not create them yourself in your MyExercises chapter folders.

Explore ArcMap Menu bar

1   On the ArcMap Menu bar, click File. Many of the familiar Microsoft-style commands, such as Save and Print, are available on this Menu bar, along with several specialized commands for GIS.

2 From the File menu, click Map Document Properties and make sure the "Store relative path names to data sources" check box is selected. There are two critical points to make here: (1) an ArcGIS map document, such as Tutorial2-1.mxd, does not store any of the map layers displayed, such as Rivers or Tracts (these data sources can be located anywhere on your computer, computer network, or even on the Internet, and a map document can just point to them and use them from there); and (2) if your data sources are on your computer, it is important to take the relative-paths option. That way, the pointers to data sources will still work if your instructor or someone else places a copy of your work on a different hard drive or path, as long as the folders and files used remain in the same relative position. So, make it a rule to use this selection. (It is not the default, so you have to make the selection.)

3 Click OK.

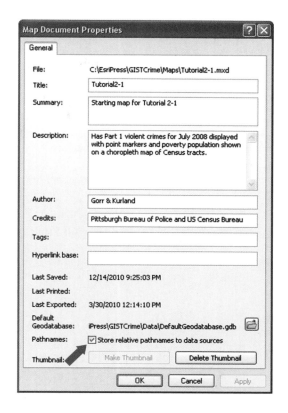

YOUR TURN **YOUR TURN** YOUR TURN YOUR TURN YOUR

Open each menu on the ArcMap Menu bar and familiarize yourself with what is available on each drop-down list. Do not, however, make any selections. If you do, you might make changes in the map document that will make it difficult to follow subsequent exercises.

Explore the table of contents

One of the most important parts of the ArcGIS interface is the table of contents—the panel that displays map layer names and symbols, similar to a map legend. The table of contents is more than just a display. It is also a part of the interface with which you interact.

1 In the Table Of Contents window, make sure the List By Drawing Order button is selected. Press your mouse button on the Tracts: Poverty Quantiles layer near the bottom of the table of contents and drag that layer to the top of the list, just below the word Layers. Then release it. If the table of contents is not visible in ArcMap, click Windows on the Menu bar, and then click Table Of Contents. ArcMap draws from the bottom up in the table of contents. So now, ArcMap draws tracts last, or on top of the other layers. Because the Tracts layer has color fill for areas that cover all of Pittsburgh, it now covers all other map layers below it in the table of contents.

2 On the Standard toolbar, click the Undo button . Tracts returns to the bottom of the table of contents, so that ArcMap draws it first and draws smaller features on top of it, thus making them visible.

3   **Select the check box to the left of Commercial Areas to turn that layer on.** The Commercial Areas layer is a buffer created in ArcGIS that consists of a 600 ft radius, about the length of two city blocks, around areas of Pittsburgh that are zoned for commercial use. Quite often, pedestrian traffic to and from commercial areas spills over into immediately surrounding areas—hence, the buffer depicting this spillover. The Commercial Areas layer has a crosshatch that allows you to see the Tracts layer underneath it. You can turn Tracts off to see the commercial areas more clearly, but then turn Tracts on again. You can see that a lot of serious violent crimes occur in poor and commercial areas. There is a lot of human interaction and poor informal guardianship in these areas, so criminal perpetrators can easily mix in with the crowd without raising suspicions. Commercial areas also provide many good targets for crime.

Internet keyword search   **routine-activity crime theory**

4   **Turn the Commercial Areas and Tracts: Poverty Quantiles layers off, and turn Streets on.** You can see now why a lot of areas do not have crimes: they do not have streets, and so are not frequented areas. Such areas include cemeteries, reservoirs, parks, and (in the case of Pittsburgh) uninhabitable steep hillsides. If you are building a crime map for a jurisdiction, you can often obtain separate map layers for such areas from the local government. Pittsburgh has an "island" municipality, Mount Oliver, which lies entirely within city boundaries. The Streets layer, which does not include Mount Oliver streets, has a gold color right now, but you can change it to more of a background color, such as light gray.

5   **Click the gold line below Streets in the table of contents.**

6   **In the Symbol Selector window, click the Color arrow, select the fourth paint chip down in the first column (Gray 30%), and click OK.** There is a valuable cartography lesson here: use shades of gray for map layers such as streets that are there to provide spatial context. Use bright colors only for the subject of interest, such as crime points. As an exception to using gray for spatial context, use blue for water features—making such selections gives the map a more natural appearance. The result will be a professional-looking map composition in which the relevant information such as crime points "jumps out" at you.

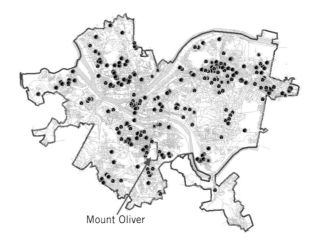

Mount Oliver

7   **Right-click Streets in the table of contents and select Open Attribute Table.** Take a look at the attributes in this table. The format of street files is studied in detail in chapter 9, but notice for now that there are street numbers for only the left and right sides of the ends of street segments, which generally are a block long—see FRADDL (from address left) through TOADDR (to address right) in the column headings. From (FR) and To (TO) end points of a street segment are determined by the order of digitizing points, with a From point being the first point digitized on a line. This is a limitation of common street maps that reaps consequences on the positional

accuracy of map locations for street addresses. Street addresses are approximate locations on a map created through simple interpolation, assuming that street numbers are uniformly distributed in a block.

8 Make the attribute table longer by dragging one of its lower edges. Next, press the CTRL key, click a gray selector button to the left of a row, and drag down through the row to select about a dozen street segments. The data rows selected appear in the selection color, as do the corresponding graphic elements on the map. Take a look.

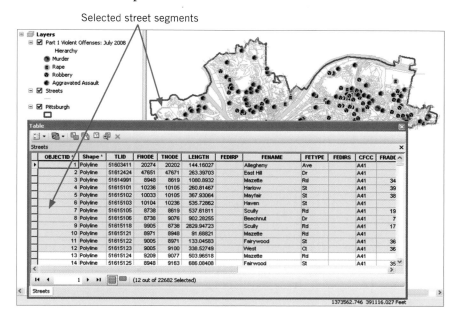

9 Close the Streets attribute table.

10 On the Menu bar, click Selection > Clear Selected Features.

YOUR TURN YOUR TURN YOUR TURN YOUR TURN YOUR

Open the attribute table for other map layers and explore the attributes. Be sure to scroll to the right in a table so you can review all the attributes available. The Part 1 Violent Offenses and Tracts layers are good layers to explore. Try selecting records in the table and see them also show up on the map. Clear all selections when you are finished.

Explore Tools toolbar

The Tools toolbar has commonly used navigation tools and other functions. You can try out a sample in this exercise. If the toolbar is not visible, go to the Menu bar and click Customize > Toolbars > Tools. You can move the toolbar by dragging it by the bar along the top. You can also anchor the toolbar to another menu or an edge of the map by dragging the toolbar there. To break it free again, drag its left edge when the pointer becomes a double-headed arrow.

1 **Drag the Tools toolbar to just below the Standard toolbar to dock it there.**

2 **On the Tools toolbar, click the Zoom In button** 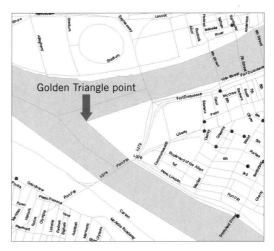 **and draw a rectangle around the "Golden Triangle" of Pittsburgh, where the three rivers converge.** The current map extent, the window that your computer screen has on the world, changes to the rectangle you created, and the map zooms in to that extent.

3 **Right-click Streets in the table of contents and select Label Features to turn street names on.**

4 **Turn street labels off by clearing the Label Features selection and click the Full Extent button** .

5 **Click the Select Elements button** **and pause over the Golden Triangle point.**

6 **Read the corresponding coordinates under the lower-right corner of the map window.** Your values should be close to 1,338,614; 411,972. These are so-called projected or flat-map coordinates in feet, using the state plane, Southern Pennsylvania zone projection. Projected map coordinates are generally quite large numbers because the unit of measure is so small—in this case, feet.

Add and remove layers in the table of contents

Map layers can easily be added and removed to and from the table of contents, although they are not physically being added or removed to and from the map. All you need is access to hard-drive storage on your computer or local area network (LAN). Note that you can have more than one copy of a map layer, and each copy can be symbolized differently for different purposes.

1 **On the Standard toolbar, click the Add Data button** **, and in the Add Data dialog box, click the Connect To Folder button** **on the Menu bar. Expand the My Computer directory by clicking the toggle key to the left. Expand the hard drive on which ArcGIS is installed (most likely C), expand the EsriPress folder, and click GISTCrime. Then click OK.** The next time you need to get to the GISTCrime folder, where all the data files needed for this book are stored, you can click the Connect To Folder button and find C:\EsriPress\GISTCrime, a direct path for use of these files. In the next step, you'll add a layer to the map.

2   **Double-click the Data folder icon > Police.gdb > Offenses2008 to add this feature layer to the map.** This feature layer has all 26 hierarchy crimes reported in the UCR for all of 2008, besides the top four already shown for July 2008. It has 42,108 points! Note that a file geodatabase (.gdb) such as Police.gdb is a folder that can have one or more feature layers, data tables, or other data objects. The file geodatabase is the preferred spatial file format for ArcGIS, but it is not the only one. There are several others.

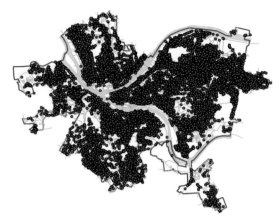

3   **Right-click Offenses2008 in the table of contents, and then click Remove.** Note that you are not actually adding or removing a *copy* of the physical Offenses2008 map layer to or from Tutorial2-1.mxd. All you are doing is adding and removing a reference to that layer in the Police file geodatabase in the Data folder where it is stored.

Add maps from the Internet

ArcGIS 10 users have built-in access to ArcGIS Online for detailed map layers. High-resolution satellite images, streets, topography, and other features are available to add to your map document. If you have an active Internet connection, do the following steps; otherwise, just read along and look at the resulting maps on the following pages.

1   **Turn Streets and Rivers off.**

2   **On the Menu bar, click File > Add Data > Add Basemap > Imagery. Click Add.** ArcGIS Online uses the map extent, or range of world coordinates, of your Pittsburgh map to extract an image mosaic, which it sends via the Internet to ArcMap for display. Note that while ArcGIS Online provides this image, it and the following image were supplied by i-cubed.

3   **On the Tools toolbar, click Zoom In and draw a rectangle around the Golden Triangle of Pittsburgh.**

4   **Zoom into the parking lot indicated by the arrow in the figure. Then right-click the image mosaic in the table of contents and click Zoom To Nearest Cache Resolution.**

As you can see, ArcGIS Online is a rich resource for crime mapping. You can get highly detailed images of crime areas that are quite small. The Zoom To Nearest Cache Resolution option provides the sharpest image available for the scale to which you zoomed.

5   After you are done viewing the photo, remove it from the table of contents.

6   Turn Streets and Rivers on.

7   Click the Go Back To Previous Extent button until you get back to the Golden Triangle.

N YOUR TURN   YOUR TURN   YOUR TURN   YOUR TURN   YOU

Explore more of the map layers available from ArcGIS Online. Be sure to try the Shaded Relief (topography) and the USA Topo Maps (a scanned 1:24,000 scale topographic map). Remove all the layers you add once you are done exploring them. Zoom to full extent and turn Streets and Rivers back on. When you are finished, save your work and close ArcMap.

Tutorial 2-2

Exploring the ArcCatalog user interface

ArcGIS Desktop has its own utility package called ArcCatalog, which is comparable to Windows Explorer. ArcCatalog provides many file and other utility programs that are essential to GIS. There are two versions of this package—a full-featured version that you open as a separate software package called ArcCatalog and a second version, called Catalog, that opens as a window in ArcMap to provide a selection of features from ArcCatalog. In this exercise, you open the separate ArcCatalog package first.

Use ArcCatalog to explore map layers

Note: If both ArcMap and ArcCatalog are open and you are attempting to work on a file that the other program also has in use, the files will be in contention, and you will not be able to modify the file as desired. So, be sure ArcMap is closed before starting this exercise.

1   From the Windows Start menu, click All Programs > ArcGIS > ArcCatalog 10.

2   Expand the Catalog tree in the left panel until the Data folder is visible and click Pittsburgh.gdb. Then click the toggle key to the left of Pittsburgh.gdb. Clicking the toggle key exposes the contents of the Pittsburgh file geodatabase, which consists of the several map layers seen in the right panel.

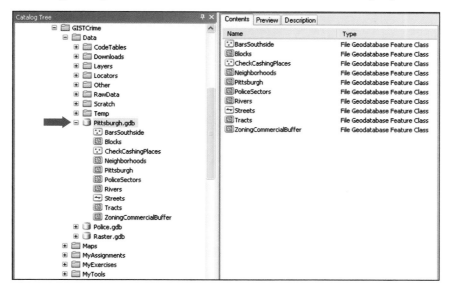

3   Click Streets under Pittsburgh.gdb in the left panel and then the Preview tab in the right panel. Make sure Geography is selected at the bottom of the window. If you were new to this map layer, the preview feature would give you a quick overview of the layer.

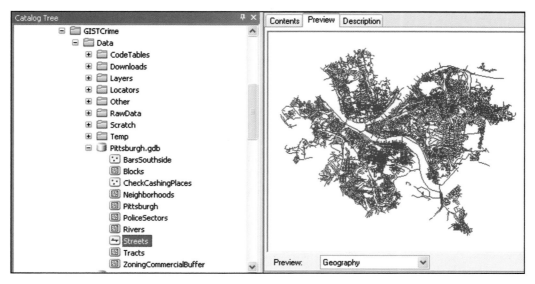

4   Click the Geography arrow, select Table, and scroll to the right in the table until you can see **FENAME** (feature name, or street name in this case) and **CFCC** attributes. CFCC (Census Feature Class Code) is a code for street type. Code A41, as seen in the preview, is an unseparated city street.

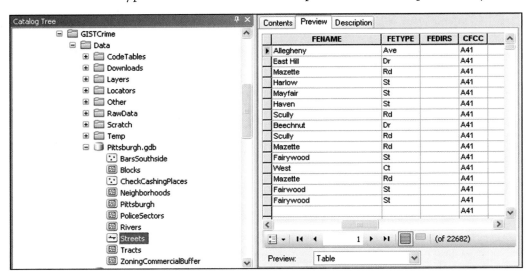

YOUR TURN   **YOUR TURN**   YOUR TURN   YOUR TURN   YOU

Metadata is "data about data." Limited metadata is available under the Description tab in ArcCatalog. Explore metadata for the Blocks layer in ArcCatalog.

Use Catalog in ArcMap to maintain map layers

A file geodatabase is a folder created in ArcCatalog or Catalog that has a name ending in ".gdb" and that stores one or more map feature classes and data tables, as you saw in the Pittsburgh.gdb folder. If you were to look in this folder using Windows Explorer, you would see many files with cryptic names such as a00000001.gdbindexes and a00000001.gdbtable. Each feature layer has several such files. Fortunately, you never have to work directly with these files, because ArcCatalog or Catalog does that for you. In this exercise, you can try out Catalog in ArcMap.

1   Close ArcCatalog and open ArcMap 10. Then open Tutorial2-1.mxd.

2   Click File > Save As. Change File Name to **Tutorial2-2.mxd** and click Save.

3   On the ArcMap Menu bar, click Windows > Catalog. Starting with Folder Connections, expand the Catalog tree to the Pittsburgh geodatabase in the Data folder. Notice the Auto Hide button 📌 in the upper-right corner of the Catalog window, which you can click to keep the Catalog window open. Alternately, you can have the window automatically collapse 🗔 when you are finished using it by clicking Auto Hide again 📌. To reopen the Catalog window, click the Catalog button on the border of the Map panel.

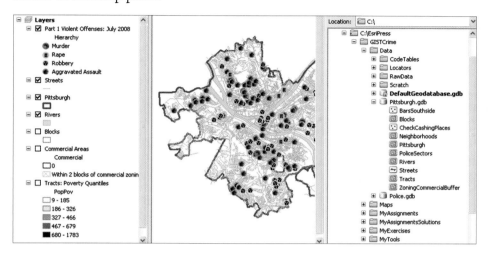

4   Right-click Rivers in the Catalog tree under the Pittsburgh geodatabase and select Copy.

5   Right-click the Pittsburgh geodatabase and select Paste. Then click OK. Catalog creates Rivers_1, a copy of Rivers.

6   In Catalog, right-click Rivers_1 and select Rename.

7   Change the name to **RiversPittsburgh**, and then click anywhere in a white area of Catalog to accept the change.

8   Right-click RiversPittsburgh, and then click Delete. Click Yes.

9   Save your map document, but do not exit ArcMap if you are continuing with the next tutorial.

Tutorial 2-3

Examining map layer properties

Each map layer has several properties you can view in ArcMap and sometimes change or enhance. Some properties are native to the stored map layer and others are related to how you wish to use the layer within a map document in ArcMap.

1 Save your map document as **Tutorial2-3.mxd** to your chapter 2 folder in MyExercises.

2 In the table of contents, right-click Part 1 Violent Offenses: July 2008 and select Properties. Then click the General tab.

3 Type the Description and Credits text as seen in the figure.

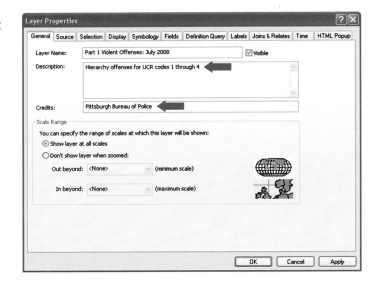

4 Click the Source tab. The Source tab shows where the map layer is stored and what projection its coordinates are in. This layer is based on the Offenses2008 feature class stored in the Police file geodatabase. (In step 8, you will see how ArcMap extracts only selected data from this feature class for display.) If you get a red exclamation point for a layer in the table of contents and the layer is not displayed, the path stored is incorrect. In that case, click Set Data Source, browse for the layer, and reset its data source property. There are many alternative coordinate systems for map layers. The coordinate system used here, state plane, is a common one used in local governments in the United States.

5 Click the Symbology tab.

6 Click the label Murder and change it to **Murder-Manslaughter**. Similarly, change Rape to **Forcible Rape**.

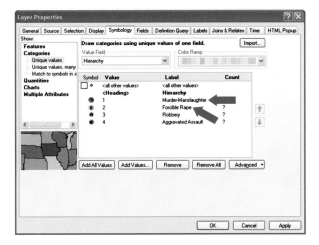

7 Double-click the orange symbol for Forcible Rape, change its size to **11**, and click OK.

8 Click the Definition Query tab. The definition query has a logical condition that selects only the desired data from Offenses2008 for display: "Hierarchy" <= 4 AND "DateOccur" >= date '2008-07-01' AND "DateOccur" <= date '2008-07-31'. While the expression syntax is new to you here, you can see nonetheless that the condition selects the points displayed on the map—UCR Part 1 crimes at hierarchy 4 or less (a lower number designates a higher crime in the hierarchy) for the month of July 2008.

9 Click OK in the Layer Properties window. ArcMap updates the legend for Part 1 Violent Offenses: July 2008 with new labels and new symbology, and also updates the map with the larger orange symbol for forcible rape.

> ☐ ☑ Part 1 Violent Offenses: July 2008
> Hierarchy
> 🔾 Murder-Manslaughter
> Ⓡ Forcible Rape
> ⊕ Robbery
> Ⓔ Aggravated Assault

YOUR TURN   **YOUR TURN**   YOUR TURN   YOUR TURN   YOUR

Explore additional properties for Part 1 Violent Offenses: July 2008—Selection, Display, Fields, and Labels. The other three properties—Joins & Relates, Time, and HTML Popup—are discussed in later chapters. Do not make any changes. Instead, just become familiar with the properties. Also, check out the properties for Tracts: Poverty Quantiles. Close Layout Properties when you are finished. Save your map document, but do not exit ArcMap if you are continuing with the next tutorial.

Tutorial 2-4

Examining Layout View

The map you've been viewing thus far is good for computer interaction, but what if you need a map for a PowerPoint presentation, a Word document, or simply a paper copy? Layout View provides an interface for assembling a map layout, including a title, legend, scale, and other map elements, which is useful for creating a stand-alone map.

1 On the ArcMap Menu bar, click View > Layout View. Until now, you have been in Data View. Now you are in the alternate Layout View.

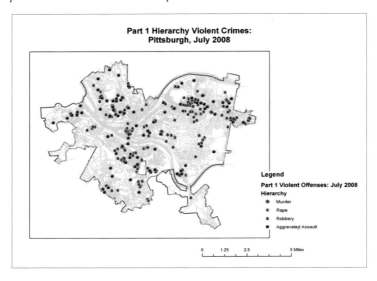

2 In the table of contents, turn Streets off and turn Tracts and Commercial Areas on. Then zoom to full extent. ArcMap dynamically updates the legend to include the tracts.

3 Click File > Export Map.

4 Click the Browse button and go to your chapter 2 folder in MyExercises for the save location, change File Name to **OffensesJuly2008**, save the type as a JPEG, make the resolution **300 dpi** (dots per inch), and click Save. The 300 dpi resolution is a publication-quality image with high resolution, so the corresponding file size is large.

5 Open a new blank Word document and press ENTER several times to open a page full of blank lines.

6 At the top of the page, type **Below is a map, created in ArcGIS Desktop 10, of serious violent crime in Pittsburgh. It shows that such crime tends to be concentrated in poor and commercial areas.**

7 Click Insert > Picture and go to your chapter 2 folder in MyExercises. Then double-click OffensesJuly2008.jpg.

Below is a map, created in ArcGIS Desktop 10, of serious violent crime in Pittsburgh. It shows that such crime tends to be concentrated in poor and commercial areas.

Part 1 Hierarchy Violent Crimes:
Pittsburgh, July 2008

8 Save your Word document as **OffensesJuly2008.docx** to your chapter 2 folder in MyExercises.

9 Save your map document and close ArcMap.

Assignment 2-1

Critique an online crime mapping system

Many police jurisdictions place crime maps—static or interactive—on the Internet as a public service. For example, you can find links to many of these Web sites on the National Institute of Justice home page by searching for State, Local & International Links. It is easy to place static map images on an Internet site, but interactive map images that let you turn layers on and off, zoom in, and interact with the map online are beyond the scope of this course. This course focuses on the ArcGIS Desktop 10 software package.

For this assignment, choose an online crime mapping site with interactive maps, and then create a summary report. Include the following in a Word document:

+ Title page, including the title, your name, and date.

+ Introduction, including the purpose of the paper and sources of information. Cite the Web site you use in a footnote (for reference style, go to the Chicago Manual of Style Online Web site and click the Chicago-Style Citation Quick Guide link).

+ Analysis section, including (as best you can determine) types of maps available (point or choropleth), two example map images (right-click an online image and select Save Picture As), input data layers, interactive map tools (for example, zoom in, identify features), components of map layouts (for example, scale bar and legend), crime types/events provided, and timeliness.

+ Summary, including your overall assessment of the site. Is there enough information so that citizens can take precautions and perhaps prevent crimes? Can block watch groups monitor crime problems in their neighborhoods?

Include the images in the analysis section as figures in report style. In text, refer to each image with a description—for example, "Figure 1 is a map of nuisance crimes in Trenton, NJ, for July 2010 showing that such crimes are highly clustered around commercial areas." Then place the image below this text and put a caption beneath it such as "Figure 1. Map of nuisance crimes in Trenton, NJ, July 2010."

What to turn in

If you are working in a classroom setting with an instructor, turn in an electronic document called **Assignment2-1YourName.docx**. Alternately, if requested, turn in a hard copy of your report.

Assignment 2-2

Compare crime maps for serious violent crimes in Pittsburgh

If you have a master crime incident file with a lot of historical data, you can easily create maps for different time periods. Tutorial2-1.mxd already has a map of serious violent crimes for July 2008.

For this assignment, export this map as an image file, create another map for August 2008 for the same crimes, and export it. Then place these images in a presentation and comment on the changes in crime patterns from July to August.

Use the following steps:

1   Open Tutorial2-1.mxd, rename it **Assignment2-2YourName.mxd**, and save it to your assignment 2-2 folder in MyAssignments. Modify the map properties to correspond to this assignment.

2   Right-click Part 1 Violent Offenses: July 2008 in the table of contents and click Copy. On the Menu bar, click Edit > Paste to add a second copy of this layer to the table of contents.

3   Modify one of the copies so that its crimes are for August 2008. Keep the same crime types. **Hint:** Modify the Query Definition property shown in the exercises in this chapter. In the query, change the month number from 07 to **08**.

4   Turn Tracts and Commercial Areas off and turn Streets on. Make sure the streets are a gray color.

5   Export two maps in Layout View in JPEG format with **150 dpi** resolution, one called **July2008.jpg** and the other **August2008.jpg** for the corresponding months of crimes. Do July first. Then change the month in the map layout title to August 2008. Double-click the title to change it.

What to turn in

Create a PowerPoint presentation called **Assignment2-2YourName.pptx**. Include a title slide with the title and your name. Import the two images as slides, positioned exactly the same for each slide, so you can click back and forth between them to see the differences. Include a slide that describes some differences between the two months for aggravated assaults. **Hint:** Describe differences separately for each of the three areas created by the three rivers in Pittsburgh. Call them **Northwest**, **Southwest**, and **East**. Also turn in **Assignment2-2YourName.mxd**.

OBJECTIVES

Use maps designed for the public
Use an early-warning system map
Use a pin map for field officers

Chapter 3

Using crime maps

Different audiences or users need different kinds of information. In this chapter, you learn how to use sample crime maps for three user types: (1) field officers, (2) investigators and top management, and (3) the public. Map use includes navigation, information retrieval, and output. In chapter 4, you learn how to build the same maps from scratch, and in chapter 10, you learn how to automate map production to include periodically updated crime data.

Crime maps and their uses

Chapter 2 gives you a head start on using ArcGIS by working with some of the features and tools that are part of the user interface. This chapter takes you further down the path to becoming a crime mapping and analysis expert by having you sample maps that are derived from the system you will be building and using in the rest of this book. Users of crime maps fall into different groups, and each group has its own needs. This chapter starts with a discussion of internal user groups—field officers and investigators—and proceeds to maps for the public, thus meeting a police organization's highest priority needs first. The tutorials, on the other hand, follow a different order, starting with a public crime map, because it is the simplest to use, and progressing to the more complicated crime maps used by field officers and investigators.

Crime maps for field officers

As the primary agents of law enforcement and crime prevention, field officers on patrol or walking a beat are experts on their assigned police sectors: they know the streets, the buildings, and the people who frequent them. It is not necessary to provide a lot of spatial context in map compositions for these users. Thus, the maps you use in this chapter have the minimum number of layers (streets, rivers, and police sector boundaries). In practice, you can always add more feature layers if needed to provide spatial context.

A major benefit of crime mapping done for field officers is the integration of crime data—filling in the missing pieces that identify or complete crime patterns in the making. Field officers know the crimes that occur in their sector when they are on duty, but missing from the total picture may be crimes that occurred on other shifts or on their days off, as well as crimes that occur in nearby police sectors that spill over into their sector. The solution is to provide traditional "pin maps" of crime points that are augmented by real-time, comprehensive data through the use of GIS. Before GIS was available, pin maps were large wall maps with pins literally stuck in them to mark crime points, color-coded by crime type. Obviously, access to such maps was limited, and the maps were difficult to keep up to date. The use of GIS in crime mapping changes that.

Along with the integration of crime data, crime maps for field officers can also illustrate a persistence or change in crime patterns. Thus, it is important that pin maps for field officers provide *temporal crime context* as well as spatial context. The map you use in this chapter has a four-week moving window of data (28 days ending with the current day). The most recent several days (for example, three to seven days) have large, bright-colored point markers representing crimes by crime type, while the balance of older days use smaller, paler versions of the same point markers. Thus, map users can determine whether spatial clusters of crimes have abated, are continuing, or are new crimes. For example, in the figure map, you can see emerging larceny "hot spots," places where more larceny is occurring, and a large larceny hot spot that is declining.

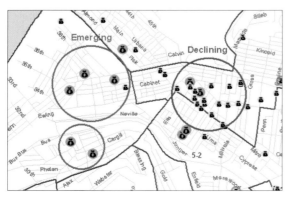

Larceny hot spot dynamics from a four-week time window: weeks 1–3 with smaller point markers and week 4 (most recent week) with larger point markers.

The use of ArcMap navigation tools allows for more effective use of pin maps. Field officers need to be able to quickly access a police sector, which can be done with the use of spatial bookmarks, and perhaps zoom in on a particular area to get a closer view. ArcMap also has a feature that automatically produces a series of map images for areas such as police sectors. It is easy to build a low-cost intranet system to deliver such images on a real-time basis. Having a map at hand, an officer can then access the data records for selected points to get the details on crime offenses—a task that GIS makes easy, given the linkage between graphic features on the map and corresponding attribute records presented in the form of tables.

Early-warning system maps for investigators and top management

The crime mapping needs of investigators and top management differ from the rank-and-file, because, unlike field officers, they do not have individual territories of responsibility and the detailed knowledge that goes along with that smaller scale. Instead, investigators and management must scan an entire jurisdiction for potential problems, sometimes with limited personal contextual knowledge of the entire breadth of a jurisdiction. This scenario calls for an early-warning system, in which color-shaded choropleth maps draw attention to police sectors or other areas that have potential problems (the scan), and then allow the user to drill down to specific points for detailed diagnosis of selected areas (the analysis). Like the field officers' pin map, the early-warning system depends on the integration of crime data with both spatial context that orients the user and temporal context that differentiates established crime patterns versus crimes that are just emerging. All these components are part of the system you use in this chapter.

Members of this crime mapping audience include major-crimes investigators (narcotics, burglary, violent crimes, and other squads) as well as top management. Of course, the crime types mapped vary by squad, whereby the burglary squad, for instance, gets maps of burglary and related crimes. Top management has responsibilities that stretch across a jurisdiction, including the accountability of decentralized units as well as setting priorities for the entire department. At regular CompStat (computer statistics) meetings, precinct or zone commanders decide how to allocate their police resources to solve crime problems while remaining accountable through the review of performance measures such as crime levels, officer overtime expenses, and citizen complaints of police abuses. Such meetings often use early-warning system maps to establish management objectives, while leaving the crimefighting strategy to individual field units, which have the detailed expertise to solve local problems (and can use pin maps to drill down to the needed information). CompStat meetings also provide a forward-looking perspective by determining the best targets for crime prevention and law enforcement. This sort of crime map analysis has acquired the label "predictive policing" for its goal of predicting the next targets. A related form of community policing, SARA (scanning, analysis, response, and assessment) policing, is also used for problem solving.

Internet keyword search   **CompStat, predictive policing, SARA policing**

In this chapter, you use two choropleth maps for the current month that make up an early-warning system for burglary investigators. One map displays the level or number of burglaries per police sector for the current month. Police sectors that have relatively high burglary levels are obvious candidates for crime suppression and prevention measures. The second choropleth map displays the change in the level (or number) of burglaries per police sector—the difference between crime levels in the current month and the previous month. Police sectors that have recently experienced large crime increases are also candidates for crime suppression and prevention.

After the map user identifies police sectors needing additional police attention, additional information is needed for further diagnosis of crime problems. A pin map showing selected crime types for the most recent month can provide such information. Research by Bowers, Johnson, and Pease (2004) demonstrates that repeat burglary locations and surrounding locations are prone to future burglaries, and thus are good targets for directed patrol. Cohen et al. (2007) and Gorr (2009) provide research demonstrating that leading-indicator crimes exist for serious crimes such as burglaries and that the leading indicators tend to increase before the number of burglaries increases. Such leading indicators are thus predictive of future increases in burglaries. The pin maps that make up the early-warning system include burglaries and two leading indicators of burglaries—vagrancy and disorderly conduct crimes. For these maps, you will be using size-graduated point markers to show locations that had repeat burglaries in July—the more burglaries at the same location in the same month, the larger the point marker. The figure shows a change map for burglaries for July 2008 with the ArcMap Magnifier window drilling down to the police sector in Pittsburgh that had the highest increase in burglaries—Police Sector 5-7.

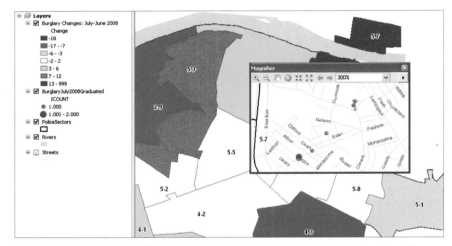

In the Magnifier window of this change map, you can see that one location, the point with the larger point marker, had two burglaries in July.

Early-warning system maps need sophisticated forms of map navigation. These include map layers that automatically turn on and off depending on the map scale, or how far in or out the user zooms in. Other valuable navigation tools include the Magnifier window and Overview map, which allow the user to simultaneously view maps at different scales. The map user may also wish to obtain and compare statistics or make measurements, and the ArcMap tools to do these functions are discussed later in the chapter.

Crime maps for the public

While there are additional clients for crime maps—for example, the news media and academic researchers—the final audience considered in this section is the public (**Note:** Assignment 3-2 is to build a map for the news media). There are considerations for creating maps for the public that go beyond those for police use. First, maps produced for the public need to stand alone, and thus contain additional map layout and interpretation elements. The solution is to place maps in layout format, including a descriptive title, legend, scale, and other map-reading aids. It is easy to export such map layouts as image files for use in reports, presentations, or on the Internet for use by the public.

Second, it is sometimes necessary to protect the privacy of individuals, such as victims, by disguising map locations. Hence, maps for the public call for some degree of error to be added to point locations while still conveying enough information to aid in crime prevention. The approach used in this book is to move crime locations from the street address of occurrence to random locations within the same city block of occurrence—the approach used by the Washington, D.C., Metropolitan Police Department. In the accompanying map, you can clearly see problem areas—for example, burglaries in the north part of the Pittsburgh neighborhood—although the exact locations have been changed.

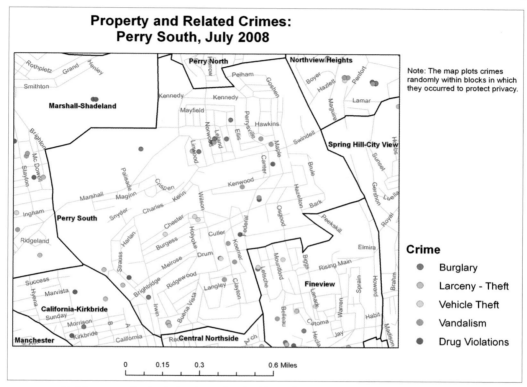

Public map of selected crimes for the Perry South neighborhood in Pittsburgh.

Using crime mapping and analysis system maps

In this tutorial, you learn how to interactively use finished crime maps in ArcMap. ArcMap provides many tools and options for interactive map use. First are map navigation tools for finding and moving to areas of interest on the map. Second are zoom-in tools for drilling down to detailed map layers and attribute data for analysis once you are in an area of interest.

Tutorial 3-1

Using maps designed for the public

Using maps designed for the public is simple and therefore a good starting point for the exercises in this chapter. Public users need not be direct users of GIS. Instead, they will more likely be using a Web site and hyperlinks to navigate to images of map layouts that are output by GIS crime mappers.

Open map document

1   From the Windows Start menu, click All Programs > ArcGIS > ArcMap 10.

2   In the ArcMap window, click Existing Maps, and then click the Browse button ⬚ to search for more.

3   Go to the location of the Maps folder (for example, C:\EsriPress\GISTCrime\Maps if it has been installed on the C drive), select Tutorial3-1.mxd, and click Open. The map layout that is generated, from the same map composition as in the preceding figure, shows property offense crimes and is zoomed to the Shadyside neighborhood of Pittsburgh. A map layout such as this is the final product you would export for use in a document, presentation, or on a Web site.

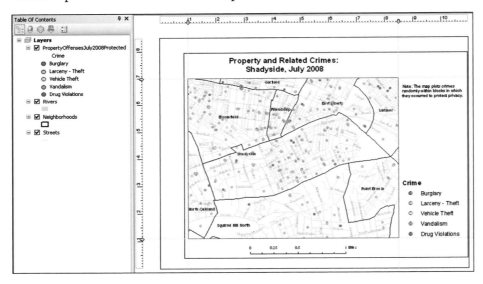

4   Click File > Save As. Go to your chapter 3 folder in MyExercises and click Save.

Use fixed zoom

The displayed area of the map is called the map extent. It is the viewing window you have of the total map and is defined by the map coordinates of any two diagonally opposite corners of the window. Zooming in a fixed distance on the center of the current display changes the extent, increasing the map scale, and thus showing more details of the Shadyside neighborhood. You can see the current scale, which depends on the physical size of your computer monitor, in the white field on the Standard toolbar at the top of the ArcMap window. It has a scale fraction such as 1:19,995 (your scale will be different), indicating that 1 inch on the screen is equal to 19,995 inches on the ground. When you zoom in, the scale fraction gets larger, so the scale is larger. Zooming out a fixed distance will show neighborhoods surrounding the Shadyside neighborhood at a smaller scale.

1   If the Tools toolbar is not visible (check to see if it is anchored above the map panel), go to the Menu bar and click View > Customize > Toolbars > Tools.

2   On the Tools toolbar, click the Fixed Zoom In button ⚏ three times.

3   On the Tools toolbar, click the Fixed Zoom Out button ⛶ three times.

Shift display using the Pan tool

Panning shifts the display to the left, to the right, up, or down without changing the current map scale and allows you to move to another neighborhood in the city.

1   On the Tools toolbar, click the Pan button 🖑.

2   Move the cursor anywhere on the map.

3   Press and drag the mouse pointer toward the bottom of the map, and then release it. The map moves upward, sending the display south, or panning to the south.

4   Pan south until you see the Greenfield neighborhood.

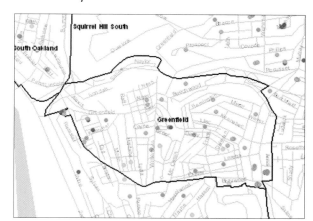

Zoom in and out using a window

Zooming in by drawing a window around an area allows you to enlarge the map to see the Greenfield neighborhood better.

1   On the Tools toolbar, click the Zoom In button.

2   Press and hold the mouse button on a point above and to the left of the Greenfield neighborhood.

3   Drag the mouse pointer below and to the right of the neighborhood and release it.

4   On the Tools toolbar, click the Zoom Out button 🔍. Zooming out allows you to pick any point on the map so you can zoom out around it.

5   Click on the left side of the Greenfield neighborhood to zoom out from the point you pick.

6   Click the Go Back To Previous Extent button. Clicking this button returns your map to its previous extent.

7   On the Tools toolbar, click the Go To Next Extent button ➡. Clicking Go To Next Extent moves the view forward in the sequence of zoomed extents you have already viewed.

8   On the Tools toolbar, click Full Extent. Clicking this button shows the full available map extent—in this case, all of Pittsburgh.

Use the Zoom and Pan functions to observe property crimes in other Pittsburgh neighborhoods.

Find features

The Find tool is used to locate features in a layer or layers based on their attribute values. You can then use this tool to select, flash, zoom, bookmark, identify, or clear the selection of the feature in question.

1   On the Tools toolbar, click the Find button 🔍.

2   Click the Features tab, type **Hazelwood** in the Find window, select Neighborhoods in the "In field" directly below the Find field, and then click Find. The value Hazelwood will appear in the bottom panel of the Find dialog box.

3   Right-click Hazelwood in the bottom panel and select Zoom To. ArcMap zooms to and briefly highlights the Hazelwood neighborhood.

4   Close the Find window.

Use spatial bookmarks

Spatial bookmarks save the extent of the map display or geographic location so you can return to it directly without having to use the Zoom or Pan tools. Spatial bookmarks have already been created for this map for a few neighborhoods. In these exercises, you use those first, and then create your own spatial bookmark.

1   On the ArcMap Menu bar, click Bookmarks > Perry North. Your map zooms to the Perry North neighborhood. Notice that the title does not change automatically. You will revise it later in this tutorial.

2   Click Bookmarks > Brighton Heights. Again, your map zooms to the bookmarked neighborhood.

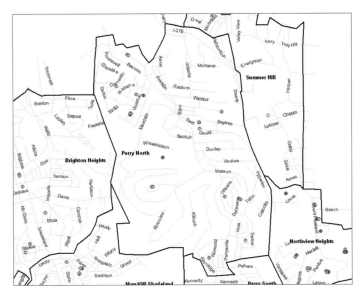

Create new bookmark

1   Use the Find tool on the Tools toolbar to move to the Greenfield neighborhood, and then use the Fixed Zoom Out tool to have it fill the map extent.

2   Click Bookmarks > Create.

3   Type **Greenfield** in the Bookmark Name field and click OK.

4   Click Bookmarks > Manage.

5   Click Greenfield, click the UP ARROW repeatedly to place the bookmark in alphabetical order, and click Close.

YOUR TURN   **YOUR TURN**   YOUR TURN   YOUR TURN   YOUR

Practice finding and creating bookmarks for the Central North Side and East Liberty neighborhoods. Place the bookmarks in alphabetical order.

Export map

Now that you have explored the map features, you are ready to export a map. First, you must change the title of the layout to reflect the zoomed area.

1   Click Bookmarks > Greenfield.

2   From the Tools toolbar, click the Select Elements tool.

3   Double-click the map title in the layout, change the neighborhood name to **Greenfield**, and click OK.

4   Click File > Export Map.

5   Change the Save location to EsriPress\
GISTCrime\MyExercises\Chapter3, choose
JPEG as the file type, change the file name to
Greenfield July 2008 Property Crimes.jpg, and
choose a resolution of 300 dpi.

6   Click Save. You can use this image in a Word
document, in a PowerPoint presentation, or on
a Web site.

7   From the Windows Start menu, click My
Computer and browse to EsriPress\GISTCrime\
MyExercises\Chapter3. Double-click Greenfield,
July 2008 Property Crimes.jpg, and after seeing
the exported image in your viewer, close the
viewer and the My Computer window.

Create a JPEG map for another of your bookmarked neighborhoods, but use a lower resolution, 50 dpi. View this image in a viewer—it's a low-quality image. When you are finished, save and close your map document.

Tutorial 3-2

Using an early-warning system map

The map you use in this exercise provides two jurisdiction-wide scans of monthly burglaries in police sectors, one for current high burglary levels (July 2008) and the other for changes from the previous month ("July minus June," or "July-June" levels). Then for police sectors of interest, which have either high levels of burglaries or large increases in the crime, the map allows you to drill down to individual crime points. To use this map, you need sophisticated map navigation tools, the ability to get attribute information from crime points of interest, and a way to highlight crime hot spots.

Open map document

1   **Open Tutorial3-2.mxd in ArcMap.** The starting display shows burglary counts for July 2008. The police sectors with the highest number of burglaries are in Precincts 3 and 5.

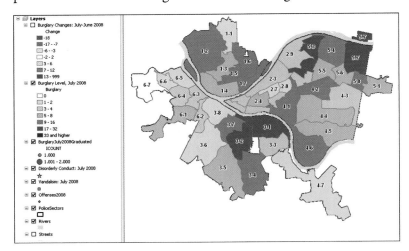

2   **Save your map document to your chapter 3 folder in MyExercises.**

Change name of a layer

Most of the layers in this map document are self-explanatory; however, you might choose to change the label of a layer. For example, a more specific label for Offenses2008 in the table of contents would be Burglaries: July 2008.

1   Right-click the Offenses2008 layer in the table of contents and select Properties.

2   Click the General tab.

3   Type **Burglaries: July 2008** as the new layer name and click OK.

YOUR TURN **YOUR TURN** YOUR TURN YOUR TURN YOUR

Change the name of BurglaryJuly2008Graduated to **Burglary Count: July 2008**. Also for the same layer in the table of contents, click ICOUNT, wait a few seconds, click it again, and delete the "I," resulting in the name COUNT.

Change layer display order

ArcMap draws layers from the bottom up. Because the Burglary Count: July 2008 layer is below the choropleth (color-coded) map in the table of contents and thus drawn underneath it, you cannot see the burglary layer. Moving it to the top of the table of contents displays it on top of the other map layers.

1   Make sure the List By Drawing Order button is selected at the top of the table of contents. You can move layers up or down in the table of contents only when this button is selected.

2   In the table of contents, press and hold the mouse button on the Burglary Count: July 2008 label and drag it to the top of the table of contents. Then release it.

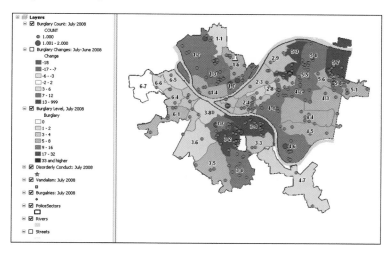

YOUR TURN **YOUR TURN** YOUR TURN YOUR TURN YOUR

Move the other three point layers to the top of the table of contents but below Burglary Count: July 2008.

Set visible scale ranges

If a layer is turned on in the table of contents, ArcMap draws it, regardless of the map scale (that is, how far you are zoomed in or out). To help you automatically display layers at an appropriate map scale, you can set a layer's visible scale range to define the range of scales at which ArcMap draws the layer, or turns it on.

1   **Turn Streets on in the table of contents.** You do not see streets yet. The map user would want to see streets when zoomed in close enough—say, to a police sector—and so would want the Burglary Level, July 2008 layer, which is covering up streets, turned off when zoomed in.

2   **Click Bookmarks > Police Sector 1-2.** Although the public generally finds neighborhoods most relevant, field officers tend to refer more to patrol district or police sector boundaries. So, this map has police sectors and sample bookmarks for them for Precinct 1. Examine the current scale on the Standard toolbar. The scale used for this map is 1:26,176, but yours will be different, depending on the size of your monitor. Whatever scale it is, it is the one you will use to switch layers on and off. In your early-warning system map, you do not want to see the details of any of the point layers until zoomed in, such as now.

3   **Right-click Burglary Level, July 2008 in the table of contents, select Properties, and click the General tab.**

4   **Click the "Don't show layer when zoomed" option. Then click the "In beyond" arrow and select Use Current Scale. Leave "Out beyond" at None. Click OK.** Nothing changes at this point. Burglary Level, July 2008 would turn off if you zoomed in any closer.

5   **On the Tools toolbar, click Zoom In and zoom in closer to a part of the current map.** Burglary Level, July 2008 turns off, its check box in the table of contents is shaded to indicate that ArcMap is using a scale range, and streets are now visible.

6   **On the Tools toolbar, click Go Back To Previous Extent.** Burglary Level, July 2008 turns back on. Next, you will implement the opposite behavior for a detailed point layer. You want Burglary Count: July 2008 to turn off when zoomed out and turn on when zoomed in.

7   **Right-click Burglary Count: July 2008 in the table of contents, select Properties, and click the General tab.**

8   **Click the "Don't show layer when zoomed" option. Make sure the "Out beyond" (minimum) scale is set to Use Current Scale and "In beyond" (maximum) scale is set to None. Click OK.**

9   **Zoom in and out to see this layer turn on and off.**

RN YOUR TURN   **YOUR TURN**   YOUR TURN   YOUR TURN   YO

Set the same minimum scale, your current scale, for the remaining three point layers—Disorderly Conduct: July 2008, Vandalism: July 2008, and Burglaries: July 2008. Use the same minimum scale for Streets. Finally, use the same scale as the maximum for Burglary Changes: July-June 2008. Try out the visible scale ranges by using bookmarks to zoom in and out of police sectors of interest. Turn Burglary Changes: July-June 2008 on and off during your exploration.

Use Magnifier window

The Magnifier window works like a magnifying glass: as you pass the window over the map display, you see a magnified view of the location under the window. Moving the window does not move or affect the current map display.

1   Zoom to full extent.

2   Click Windows > Magnifier.

3   Click the Magnifier percentage arrow and select 500%. Drag the Magnifier window over a point and release it. The magnifier scale should be one where the choropleth map turns off. If not, try a larger percentage.

4   Drag the Magnifier window until you see the crosshairs over Pittsburgh's "point" where the two rivers merge into a third river. Release the Magnifier window to get a magnified view showing crime points.

YOUR TURN   **YOUR TURN**   YOUR TURN   YOUR TURN   YOUR

Move the Magnifier window to view areas of interest. Change the magnification percentage to view smaller or larger areas. Close the Magnifier window when you are finished.

Use Overview window

The Overview window shows the full extent of the map (the City of Pittsburgh, in this case) and includes a red rectangle that shows the current zoomed extent. You can move the rectangle in the Overview window to pan the entire city while viewing the details of an area that is zoomed in.

1   Click Bookmarks > Police Sector 1-5.

2   Click Windows > Overview and drag the resulting Layers Overview window to the upper-left corner of the map panel. ArcMap displays the full extent of the map and shows a red rectangle around the current extent.

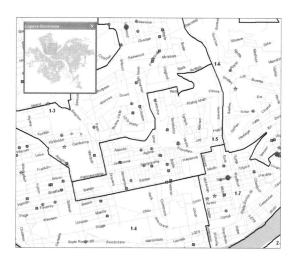

3   Right-click the title of the Overview window and select Properties.

4   Change the reference layer to **Police Sectors** and click OK.

5   Move the pointer to the center of the red
    rectangle, drag the rectangle to a new location,
    and then release it. The extent of the map
    display updates to reflect the changes made in
    the Overview window.

6   Close the Overview window.

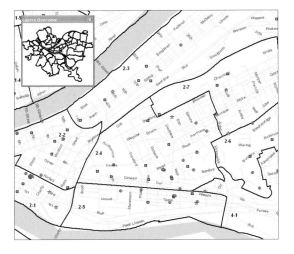

Set field properties

This task is used to prepare attribute data for use in an attribute table. You set field (attribute) properties to hide irrelevant attributes and to provide self-descriptive alias names for attributes that have cryptic names.

1   Right-click Disorderly Conduct: July 2008 in the table of contents, select Properties, and then click the Fields tab.

2   Click the Options arrow and select Show Field Aliases. A field alias is an alternate name for a field and can be used for display purposes. Initially, all aliases are the same as the field name.

3   Turn off the first eight attributes up to and including ARC_Street, and turn off Age (which is not the age of any person, but is an attribute used in chapter 9 for processing data) as well. Now the attributes that are turned off do not display when you view disorderly conduct records.

4   Change values in the Alias field in the Appearance panel for some of the remaining attributes by clicking a field in the left panel and, then in the right panel, adding spaces between words and spelling out words that are abbreviated. Here are the words spelled out for some of the more cryptic abbreviations:

    • CCN: **crime control number** (the primary police identifier for offense reports)

    • NumOff: **number of offenses**

    • ArrFName: **arrest first name**

    Click Apply and then OK when finished.

5   Right-click Disorderly Conduct: July 2008 in the table of contents and select Open Attribute Table. You can see that ArcMap uses your aliases for column headings in the table.

6   Close the attribute table.

YOUR TURN YOUR TURN YOUR TURN YOUR TURN YOUR

Turn the same attributes for Vandalism: July 2008 off and create some aliases for that layer. When you are finished, close the attribute table.

Set selectable layers

A good step in preparation for using certain tools, including the Identify tool (which you use after the following steps), is to designate which layers are selectable—that is, which layers can supply data to the Identify tool.

1 **At the top of the table of contents, click the List By Selection button .**

2 **For each of the following layers that are in the top panel of selectable layers—Police Sectors, Streets, and Rivers—click the Selectable toggle key .** Clicking the toggle key moves the layers you don't want selected to the Not Selectable panel so they won't show up in the Identify results. Should you wish to make these layers selectable within this map document, click the shaded toggle key for layers that are in Not Selectable rows.

3 **At the top of the table of contents, click List By Drawing Order.**

Use Identify tool

Use the Identify tool to display and query the data attributes of map features.

1 **Click Bookmarks > Police Sector 1-5.**

2 **On the Tools toolbar, click the Identify button , and then click the leftmost Disorderly Conduct point in Police Sector 1-7.**

3 **Click the "Identify from" arrow and choose "Selectable layers."** You can see that interactive map use is valuable for uniformed and plainclothes police officers, because they can drill down to get detailed data for crimes that are part of a spatial pattern.

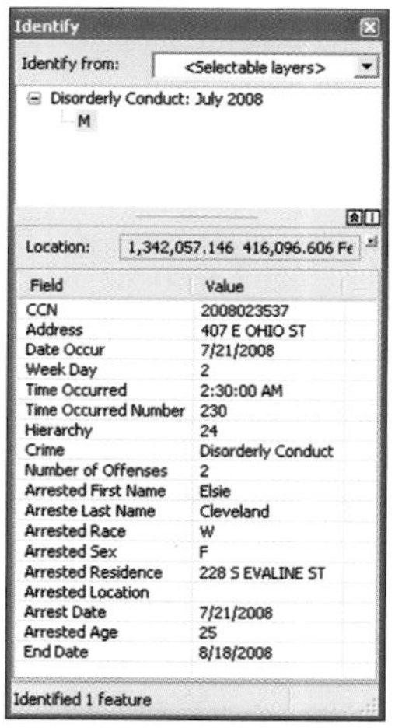

Identify a few other crimes. Try the burglary point that had two burglaries in the center portion of Police Sector 1-7, near the boundary of sector 1-6. Note that the visible layer, Burglary Count: July 2008, has two points below it in the Burglaries: July 2008 layer, both attributed to a perpetrator with the last name Clinton. Click in each of the Clinton rows to see each record. Close the Identify window when you are finished.

Measure distances

When you zoom in on a large-scale map, you can see the map details more clearly and use the Measure tool to determine the distances between crimes and other map features.

1　Click Bookmarks > Police Sector 1-2.

2　Zoom in on the small area that has three burglary points, one of which had two burglaries, in the upper-right corner of Police Sector 1-2.

3　Use the Identify tool to verify that the two burglaries at the same location happened prior to the other two, single burglary points. This suggests that the area may have been a good one for targeted patrol.

4　In the Identify window, click the records under Burglaries: July 2008 to investigate the details of an event. You can see how far away the two single-burglary locations are from the two-burglary location.

5　On the Tools toolbar, click the Measure button.

6　In the Measure window, click the Choose Units arrow and click Distance > Feet.

7　Click the location with two burglaries, and then using the street vertices as a guide, measure the shortest distance to the farther of the two other burglary points, and double-click the last point to end the measurement. It is roughly 3,200 ft between the two locations, but the measurement depends on the path you took through the network of streets.

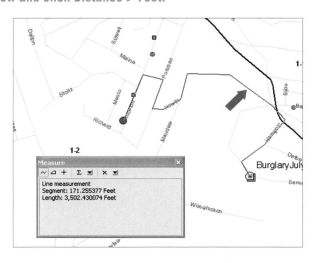

Practice measuring the distances between a few other features. Close the Measure window when you are finished, and on the Tools toolbar, click Select Elements.

Display crime hot spots

You can use the drawing tools in ArcMap to indicate potential crime hot spots on a map. Suppose the Precinct 1 sergeant would like to have patrols in Police Sector 1-6 focus on the several burglaries and vandalism points in the northern portion of that sector. You can draw a circle around these points to indicate a hot spot.

1   Click Bookmarks > Police Sector 1-6 and zoom closer into that sector so that streets and crime points are displayed.

2   From the Menu bar, click Customize > Toolbars > Draw. Using this option adds the Draw toolbar to the ArcMap interface. You can let the toolbar float or dock it above the map.

3   From the Draw toolbar, click the Draw Shapes arrow (fourth button from the left on the toolbar) and select Circle.

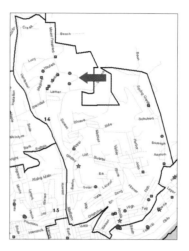

4   Click in the center of a cluster of crime points and drag the tool to form a circle encompassing the cluster of crimes.

5   Double-click the circle graphic you drew and click the Symbol tab.

6   Choose "No color" for the fill and a bright red color with a width of 2 for the outline.

7   Click OK.

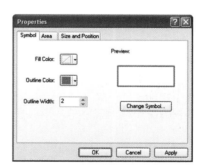

8   Move your circle, if necessary, by dragging it to position it better around your burglary points.

9   Close the Draw toolbar and save your map document.

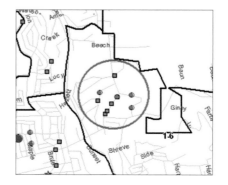

Tutorial 3-3

Using a pin map for field officers

Field officers need detailed data for crime incidents that occur in or around the police sector they patrol. It is important for them to be able to select map features for analysis and to be able to access additional information through documents you hyperlink to crime points and crime statistics you create with the use of ArcMap. Using the ArcGIS Select tools along with the attribute tables for GIS features can yield powerful results that help field officers know more about the crimes occurring in their sector. In this tutorial, you learn how to use a pin map.

Open map document

1   **Open Tutorial3-3.mxd in ArcMap.** This map shows serious crimes in the city of Pittsburgh during a four-week period, ending July 11, 2008. One layer, with smaller point symbols, shows crimes for the first three weeks, and the other layer shows crimes for the week ending July 11.

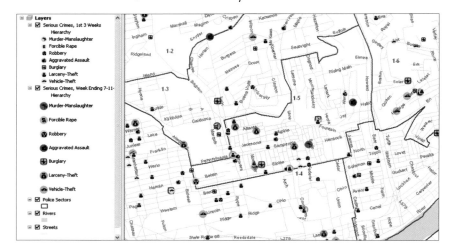

2   **Save your map document to your chapter 3 folder in MyExercises.**

Select features

There are several ways to select features. One of them is through direct selection on the map document.

1   **On the Tools toolbar, click the Select Features tool** 🔲.

2   **Click inside Police Sector 1-5, but not on a crime point marker.** ArcMap selects the police sector as indicated with a thick, bright blue line.

Change selection color

You can change the selection color to better see selected features.

1   From the ArcMap Menu bar, click Selection > Selection Options.

2   Click the Color arrow in the Selection Tools
    Settings panel and select Mars Red (second
    column, third row).

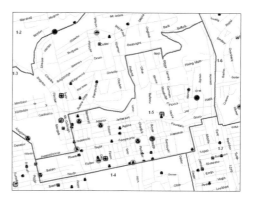

3   Click OK.

Zoom to selected features

1   On the Tools toolbar, click Full Extent. This
    map is meant to be viewed while zoomed to
    relatively small areas. At full extent, there is
    just a jumble of point markers, as you can see.

2   In the table of contents, right-click Police
    Sectors and click Selection > Zoom To Selected
    Features. That's better. Now the symbols
    provide meaningful information.

Work with attribute tables

Selected features on a map, such as crime points, are also selected in the feature attribute table, so it
makes the attribute table a good place to observe crime details.

1   In the table of contents, turn the Serious Crimes,
    1st Three Weeks layer off. Then press SHIFT
    and use the Select Features tool from the Tools
    toolbar to select all five Larceny-Theft locations
    for Serious Crimes, Week Ending 7-11-08 in
    Police Sector 1-5.

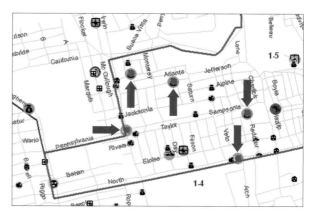

2   In the table of contents, right-click the Serious Crimes, Week Ending 7-11-08 layer and select Open
    Attribute Table.

3   At the bottom of the attribute table, click the "Show selected records" button ▤. Although you selected only five points, six crimes are selected in the feature attribute table. That is because two larcenies happened at the same address, 802 Pennsylvania Ave.

OBJECTID_1 *	Shape *	STATUS	SCORE	MATCH_TYPE	SIDE	MATCH_ADDR
21632	Point	M	100	A	L	513 ARMANDALE ST
21769	Point	M	100	A	L	1331 REDDOUR ST
21792	Point	M	100	A	R	802 PENNSYLVANIA AVE
21833	Point	M	100	A	L	1703 BUENA VISTA ST
21967	Point	M	100	A	R	802 PENNSYLVANIA AVE
21979	Point	M	77	A	L	321 W NORTH AVE

4   At the top of the attribute table, click the Clear Selection button ⊠. Then at the bottom of the table, click the "Show all records" button ▤ and close the table.

YOUR TURN   YOUR TURN   YOUR TURN   YOUR TURN   YOU

Select the other serious crimes in the map for Police Sector 1-5. Open the attribute table and observe the details for these crimes. When you are finished, clear the selection, click "Show all records," and close the attribute table. Then, turn the Serious Crimes, 1st Three Weeks layer on.

Sort fields in an attribute table

1   In the table of contents, right-click the Serious Crimes, Week Ending 7-11-08 layer and select Open Attribute Table.

2   Right-click the column heading DateOccur and select Sort Ascending. An ascending sort sorts crimes starting with those that occurred at the beginning of the week and progressing to those that occurred at the end of the week.

CCN	Address	DateOccur	WeekDay	TimeOccur
2008021284	337 ARABELLA ST	7/5/2008	7	1:40:00 AM
2008021293	BRUSHTON AV & FINANCE ST	7/5/2008	7	3:34:00 AM
2008021307	2548 CHAUNCEY	7/5/2008	7	10:10:00 AM
2008021324	5606 AYLESBORO ST	7/5/2008	7	12:20:00 PM
2008021328	305 6TH AV	7/5/2008	7	12:50:00 PM
2008021330		7/5/2008	7	12:50:00 PM

3   Right-click the column heading Address, select Advanced Sorting, and make selections as shown in the figure.

4   Click OK.

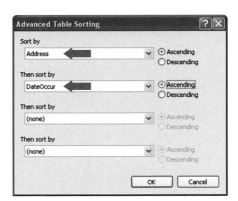

5   Scroll down in the table to 1111 Brabec St. Advanced sorting sorts crimes by both the address and the date that serious crimes occurred, making it possible to find addresses with repeat crimes, such as the two burglaries at 1111 Brabec St.

2008021653	1111 BRABEC ST	7/7/2008		2	7:20:00 AM		5	Burglary
2008021826	1111 BRABEC ST	7/8/2008		3	11:30:00 AM		5	Burglary

Move fields

Moving fields in the table affects the table display only. It does not physically move columns in the stored data.

1   Click in the column heading ArrAge, drag it to the left, and release it between ArrSex and ArrFName.

2   Similarly, move ArrRace to the right of ArrSex.

Select attribute table records

1   Sort ArrAge in ascending order.

2   Scroll down near the bottom of the table until you come to the first row with an age greater than 0.

3   Press SHIFT, click the gray selector box to the far left for the first record with AgeArr greater than 0, scroll down to the bottom of the table, and click the gray selector box for the last record (to select all records with AgeArr greater than 0). You should have 48 of 416 rows selected, showing all serious crimes that have an arrest attribute such as age for the arrested person. These records also appear in the selection color on the map (for example, the homicide in sector 1-5).

Get statistics

You can get statistics on the age of arrested persons or on any other attributes of interest from the table. If you have records selected, such as is the case now, you'll get the related statistics for only those records.

1   Right-click ArrAge and select Statistics. The age of persons arrested for serious crimes ranges from 12 to 56 with an average age of 25 rounded to whole years. The bar chart distribution is interesting, showing two modes (or relative maximums on the bar chart)—one for the late teens and a smaller one for the early 40s.

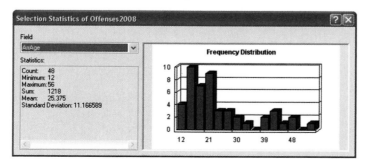

2   Close the Statistics and Table windows. From the Menu bar, click Selection > Clear Selected Features.

Add hyperlinks

It is possible to link one or more files of any kind to mapped features. Suppose, for instance, the crime mapping unit links narrative data and other files to all homicides. So, for practice, link a narrative and image files to the homicide (murder-manslaughter) in sector 1-5.

1 **From the Tools toolbar, click the Identify tool, and then click the center of the homicide point marker in Police Sector 1-5.** You will need the indicated identifier, Christian, in step 2.

2 **Right-click the identifier, Christian, in the Identify window.**

3 **Select Add Hyperlink and click Browse.**

4 **Go to Data > RawData and select Narrative2008021985.doc. Click Open > OK.**

5 **Repeat steps 3 and 4, but select Gun20080221985.jpg as the file to hyperlink. Close the Identify window when you are finished.**

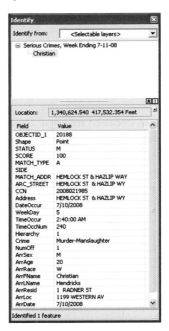

Use hyperlinks

1 **From the Tools toolbar, click the Hyperlink tool** [image]. Any point that has hyperlinked content, such as the homicide in sector 1-5, has a blue dot added to its center.

2 **Click the blue dot of the homicide in sector 1-5 with the tip of the lightning bolt of the tool.**

3 **In the resulting window, click in the row that contains the Word document and click Jump.** The (fictional) narrative document for the homicide opens in Microsoft Word.

> (Fictional document)
>
> **Narrative on 20080221985:**
>
> Witnesses saw Christian Hendricks shoot the victim, Bill Jones, at the corner of Hemlock St. and Hazlip Way at 2:40 a.m. on 7/10/2008. Both parties had been arguing at an illegal bar and stepped outside. Hendricks pulled out a handgun, a 9mm Taurus, and shot Jones three times. Hendricks fled the scene but was later apprehended in his apartment.

4 **Close the Word document, click in the row in the Hyperlinks window with the JPEG, and click Jump.** An image of a pistol opens in a viewer if you have a default viewer associated with image files.

From Shutterstock, courtesy of Billy Hoiler.

5 **Close your image viewer and close the Hyperlinks window.**

6 **Save your map document and close ArcMap.**

Analyze hot spots for larceny crimes

Larceny is among the most frequent serious property crimes, so it generates a lot of work for police and is a priority concern for the public.

For this assignment, modify and use the field officers' pin map for hot spot analysis. Of interest are relatively small areas that field officers can target for increased patrol, apprehension of larcenists, and crime prevention. More specifically, areas of interest are hot spots that appear to be extinguished, are persisting, or are emerging. Specific criteria are as follows:

- *Extinguished hot spots*—had no new larcenies in the last week but had at least six larcenies in the prior three weeks
- *Persisting hot spots*—had two or more larcenies last week and six or more larcenies in the prior three weeks
- *Emerging hot spots*—had three or more larcenies last week and, at most, one larceny in the prior three weeks

You will need to find one example of each kind of hot spot. Also, display the hot spots as follows:

- At a map scale of 1:5,000 with ArcGIS at full screen on your computer
- Identified with circles no larger than 1,500 ft in diameter (about 20 to 25 city blocks)

Instructions

1 Open Tutorial3-3.mxd and save the map document as **Assignment3-1YourName.mxd** to your assignment 3-1 folder in MyAssignments.

    a. For Serious Crimes, 1st 3 Weeks and Serious Crimes, Week Ending 7 11 08, use the Symbology tab of the Properties window and remove all crime types except Larceny-Theft. (Click in a crime row in that window, and then click Remove.)

    b. Rename these two layers in the table of contents, replacing Serious Crimes with **Larceny-Theft**.

    c. Zoom to full extent and zoom in to find hot spots.

2 Draw circles around each hot spot.

    a. Use the Draw toolbar to draw a circle around each of the three hot spots you identify. Use No Color for the fill and a shade of green for the outline.

    b. Convert each circle into a shapefile (**ExtinguishedHotSpot.shp, PersistingHotSpot.shp**, and **EmergingHotSpot.shp**) saved to the assignment 3-1 folder. Select a drawing by clicking it, click the Drawing arrow on the Draw toolbar, and select Convert Graphics To Features. Remove the circles you drew, leaving the shapefile displayed. Notice that street names do not appear inside the circles you draw, although they do with the shapefile equivalents.

3   Annotate each hot spot. ArcMap automatically creates an attribute called Name for your hot spot shapefiles. Enter a descriptive title for Name, but you will need to be in Edit mode for that purpose.

     a.   From the Menu bar, click Customize > Toolbars > Editor.

     b.   Click Editor > Start Editing. Then click the name of your hot spot. Click OK.

     c.   Open the attribute table for your hot spot, increase the width of the Name attribute, type a name such as **Emerging hot spot, 7/11/2008**, and click in the table but outside the Name field. Click Editor > Stop Editing. Click Yes.

     d.   Open the Properties window for your hot spot shapefile, click the Labels tab, and select the "Label features" check box. Modify the text symbol to your liking.

     e.   Open an Overview window for each hot spot, using Police Sectors as the reference layer.

4   Build a PowerPoint presentation.

     a.   Make the ArcMap window active by clicking the Title bar. Use Windows commands to place the image of the active window on the clipboard by pressing ALT+PRINT SCREEN. Paste such an image on a PowerPoint slide by pressing CTRL+V. If you were to just export the map from ArcMap, the Overview window would not be included. Using PRINT SCREEN includes the Overview window.

     b.   Include a title slide with the title, your name, and the date. Include an introductory slide that lists criteria for hot spots as stated in this assignment. Include the three screen captures of your three hot spots as separate slides.

     c.   Save your PowerPoint presentation as **Assignment3-1YourName.pptx** to your assignment 3-1 folder.

What to turn in

If you are working in a classroom setting with an instructor, turn in the electronic document **Assignment3-1YourName.pptx**. If requested, also turn in **Assignment3-1YourName.mxd**.

Assignment 3-2

Create maps for the media

Today's technologies change the way journalists report crimes. Some news organizations such as *USA Today* employ a database editor who does much of the organization's demographic and other data analysis. Other news organizations rely on local police departments to help editors, reporters, and broadcast journalists obtain, analyze, and present data in print and on the Web.

A local reporter needs help with a story she is doing on property offenses, particularly burglaries, in the city of Pittsburgh during July 2008. She would like to identify areas that had large increases in burglaries from June to July 2008, and then show the patterns of burglaries and other property crimes in those areas.

For this assignment, help her out by creating a map that shows the areas with large increases in burglaries and a PowerPoint presentation that explains the crime rate by police sector.

Instructions

1   Open Tutorial3-2.mxd in ArcMap.

2   Using tools such as Zoom In, Pan, and Identify, find the police sectors with an increase of 10 or more burglaries in July 2008 from June 2008. Note these as the identified police sectors. Close Tutorial3-2.mxd without saving changes.

3   Open Tutorial3-1.mxd and save the map document as **Assignment3-2YourName.mxd** to your assignment 3-2 folder in MyAssignments.

   a.   Add PoliceSectors from the Pittsburgh geodatabase and remove Neighborhoods.

   b.   Remove Drug Violations from the symbolization of crimes. (Click Properties > Symbology, click in the row for Drug Violations, and click Remove.)

   c.   Remove Drug Violations from the Layout View legend by right-clicking the legend, selecting Ungroup, and deleting Drug Violations.

   d.   Zoom in on each of the identified police sectors, one at a time, and show the property crimes for each area.

   e.   Draw one or more circles around hot spot clusters of burglaries in each identified police sector.

4   Export a layout for each identified police sector showing the property crime details for each sector.

   a.   Use the JPEG file format with 150 dpi.

   b.   Change the title of each layout image appropriately.

   c.   Include a comment with the total number of burglaries for each identified sector.
       Hint: Select crime points and examine the selected attribute records.

5 Create a PowerPoint presentation, **Assignment3-2YourName.pptx**, and save it to your assignment 3-2 folder. Include a title slide with the title, your name, and an introductory slide that explains the purpose of the presentation and the maps that follow. Import the images from your map layouts as individual slides.

What to turn in

If you are working in a classroom setting with an instructor, turn in the electronic document **Assignment3-2YourName.pptx**. If requested, also turn in **Assignment3-2YourName.mxd**.

References

Bowers, K. J., S. D. Johnson, and K. Pease. 2004. Prospective hot-spotting: The future of crime mapping? *British Journal of Criminology* 44 (5): 641–58.

Cohen, J., W. L. Gorr, and A. Olligschlaeger. 2007. Leading indicators and spatial interactions: A crime forecasting model for proactive police deployment. *Geographical Analysis* 39:105–27.

Gorr, W. L. 2009. Forecast accuracy measures for exception reporting using receiver operating characteristic curves. *International Journal of Forecasting* 25:48–61.

OBJECTIVES

Build a pin map for field officers
Build an early-warning system for investigators
Build a map for public use

Chapter 4

Designing and building crime maps

This chapter gives you the skills to design and build effective crime maps by applying good cartographic principles and the functionality of ArcMap. Much of crime mapping involves repetitive processing in which the same types of maps are produced over and over but using new data. In ArcMap, however, you can eliminate much of the repetitive work of crime mapping by saving pertinent elements such as symbols, colors, and labeling of map layers to layer files for reuse. This flexibility gives a police jurisdiction the ability to custom design maps to meet local needs and to save settings to templates for low-cost, consistent production of crime maps. As a further benefit, the resulting map templates, if combined with the macros covered in chapter 10, can fully automate the production of periodic maps. These same map design principles and methods can also be used to produce one-of-a-kind special-purpose maps when the need arises.

Map design

Designing and building maps can be satisfying and rewarding. A well-designed map provides unique information that "jumps off" the screen or the page to help the user learn more about crime patterns in a certain area and within a certain time frame. This chapter provides key map design principles and has you build state-of-the-art crime maps for use in police departments.

Periodic vs. ad hoc crime maps

Criminals are continually committing crimes and crime patterns are ever changing, so crime analysts need to produce maps on a periodic basis—by shift, by day, by week, and by month—as new data becomes available. Periodic maps help crime analysts see patterns during a given time frame, using the latest data. For example, in Pittsburgh, field officers study up-to-date pin maps of offenses in and near their sectors at roll call. These maps are produced for every roll call, every eight hours. In addition, crime analysts build special ad hoc maps that address unique crime problems as they arise—for example, a map of serial robberies showing the sequence of crimes or a map showing crimes associated with a gang rivalry.

In this chapter, you learn how to design and make maps from scratch in ArcMap. The maps you build, with the use of several interactive steps, could be used for a special one-time map or, with an additional step, they could serve as a map template for repeated use in the production of periodic maps. The additional step needed is simply to save the symbolization of selected map layers to separate layer files. Applying the symbolization of a layer file to a future map layer saves all the interactive steps of symbolizing the map layer more than once. Of course, another benefit to using such map layers is consistency: by applying the same map layer to other map layers over time, you will always get exactly the same colors, symbol sizes, and fonts for your map. Even more of a timesaver, in chapter 10, you learn how to use the ArcMap ModelBuilder application to create macros—custom computer programs that you build without writing computer code and that depend on layer files for symbolization. With ModelBuilder, you can completely automate periodic maps at the "push of a button." That way, as a crime analyst, you can spend your time creatively analyzing your periodic maps for crime patterns or quickly building special one-time maps rather than having to spend time repeating the same steps over and over to build the day-to-day maps your organization needs.

Map symbolization using attribute data

If you were using a simple drawing package to create a crime map, you'd have to choose each point marker and color and paste it in each polygon individually. ArcMap automates symbolization by using the values found in attribute tables. If a point map layer for crimes has a crime code (for example, having the values Burglary, Robbery, or Larceny), you can assign and customize a symbol for each crime code, and ArcMap does all the drawing for you. Similarly, if you have a quantitative attribute for polygons, you can choose ranges of values and assign colors to each one to show increasing magnitudes, and ArcMap will do the rest.

Cartographic principles

Cartography is the art and science of visually representing geographic features on maps with the use of symbols, lines, labels, colors, and other graphic elements to convey important information. Map design, which includes the choice of map layers as well as how they are rendered and symbolized through cartography, is based on what type of information is needed for problem solving—in this case, for law enforcement and crime prevention. If you have not read the introduction to chapter 3 on the design of maps for field officers, investigators, and top management, as well as for the public, you should do so now. It is important to understand the purpose of a map to be able to properly design it. For example, if a map is intended for study purposes and will be viewed by zooming in on small areas, you would choose larger point marker symbols than if the map were designed to be viewed at full extent.

This book is not intended to provide a comprehensive and detailed introduction to cartography and the principles of map design. Instead, you learn mostly from examples as you work through the tutorial exercises in each chapter. This book presents only the broadest of guidelines to get you started in applying GIS to crime mapping and analysis.

Much of what appears on analytical maps, whether crime maps or otherwise, is there to provide spatial context to help you discover where something of interest is located. All contextual features, therefore, should be in background (or "ground") colors, mostly grays, and the more prominent the nature of the feature, the darker the gray or the thicker the line symbol should be. For crime maps in this book, spatial context is primarily provided by streets and police sectors. Polygons, such as police sectors, should have white color fill or appear hollow with just a gray outline. Police sectors are more prominent than streets, so they should have thicker and darker gray line symbols than streets.

The subject of interest—in this case, crimes—should have bright, vibrant (or "figure") colors. If more than one kind of crime is to be shown by point markers, you would generally use different shapes *and* colors to distinguish the crimes. Use colors that are widely different to mark different types of crimes, such as blue for burglaries and red for rapes. Color fill for polygons used to represent quantitative attributes should generally use colors from the same color family—for example, shades of red, with the darker the color, the higher the level of crime. This type of color scheme is called a monochromatic color ramp. For advice on choosing colors, see Cindy Brewer's *Colorbrewer: Color Advice for Maps* on the Colorbrewer2 Web site.

The figure shows two versions of the same map: a "bad" map with misused colors for context layers and a "good" map with context layers in gray. On the bad map, it is not clear what you should be studying, but on the good map, the crime points "jump out" at you, and you can readily see the patterns. The underlying principle illustrated here is "graphic hierarchy," which you use in this chapter to direct the map reader's eye to important graphic features.

Bad map   **Good map**

"Bad" and "good" maps demonstrating graphic hierarchy.

Internet keyword search　**graphic hierarchy, figure color, ground color**

In another bad map/good map pairing (see figure), the bad map uses random colors to symbolize the population of census tracts, while the good map uses a monochromatic color ramp. Not only is the bad map hard to read and interpret, but its meaning gets lost if copied in black and white, because different colors become similar grays. The monochromatic color map, however, translates perfectly into black and white.

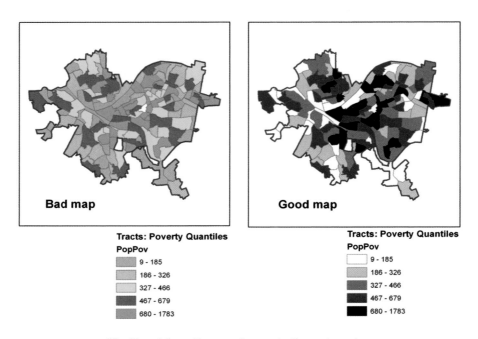

Bad map　**Good map**

Tracts: Poverty Quantiles	Tracts: Poverty Quantiles
PopPov	PopPov
9 - 185	9 - 185
186 - 326	186 - 326
327 - 466	327 - 466
467 - 679	467 - 679
680 - 1783	680 - 1783

"Bad" and "good" maps demonstrating color value.

Building crime maps

In this section, you build three maps, using prebuilt versions of the jurisdiction basemaps and crime master map layers you develop in chapters 8 and 9. The first map you build is a pin map for field officers that makes use of crime codes and sophisticated point markers that are provided by Esri for use in crime mapping.

Tutorial 4-1

Building a pin map for field officers

This first crime map uses an ArcMap function known as "unique values." Each code value of a crime type attribute gets a different, unique symbol: Murder-Homicide gets a red circle with the outline of a body; Rape gets an orange circle with the broken letter "R"; and so on. The map you build will include the first seven Part 1 crimes in the FBI crime hierarchy with violent crimes getting hot colors (red, orange, yellow, and purple) and property crimes getting cool colors (blue, green, and bluish green) to clearly distinguish the two types of crimes—violent crimes versus property crimes.

Create new map document

The basemap layers you use provide spatial context for crimes. The bright colors of the crime points make them jump out from the page while keeping the reference features, such as streets, in the background.

1   From the Windows Start menu, click All Programs > ArcGIS > ArcMap 10. Then click New Maps > Blank Map. Click OK.

2   Save your map document as **Tutorial4-1.mxd** to your chapter 4 folder in MyExercises.

3   Click File > Map Document Properties and type the following text into the appropriate sections:

   Title: **Uniformed officers' pin map**

   Summary: **Daily crime map of Part 1 crimes**

   Description: **Has 28-day window of Part 1 crimes with most recent week's crimes highlighted. Created <today's date>.**

   Author: **Your name**

   Credits: **Pittsburgh Bureau of Police and U.S. Census Bureau**

4   Select the "Store relative pathnames to data sources" check box. Click OK.

Add layers to a map document

1   On the Standard toolbar, click Add Data. Then click the Browse button and go to the Pittsburgh geodatabase. Press CTRL and click PoliceSectors, Rivers, and Streets. Then click Add. ArcMap symbolizes these layers with random colors, putting the polygon layers (PoliceSectors and Rivers) below Streets so Streets is visible. In step 4, you remove the color fill from PoliceSectors so it can go on top and not block the other layers.

2   In the table of contents, drag the layers to make the order, from the top, PoliceSectors, Rivers, and Streets.

3 Double-click the icon for each layer in the table of contents to open the Symbol Selector and symbolize as follows:

- Police Sectors: color, Hollow; outline width, **1.5**; outline color, Gray (70%)

- Rivers: color, Blue

- Streets: color, Gray (20%)

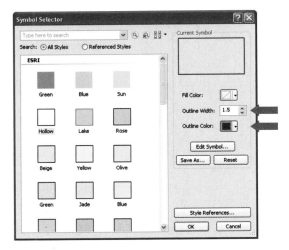

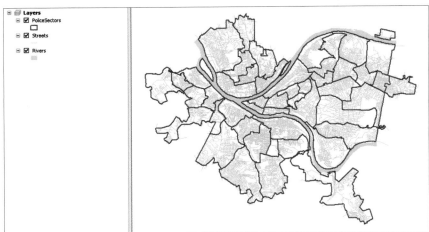

4 Double-click the PoliceSectors label in the table of contents, click the General tab, and insert a space in the layer name to make it **Police Sectors**. Click OK.

Label layers

1 On the Tools toolbar, click Zoom In and drag the tool to draw a rectangle around the northwestern part of Pittsburgh, called the North Side, to zoom in above the top rivers.

2 In the table of contents, double-click the Police Sectors label and click the Labels tab.

3   Make selections as shown in the figure, including the Bold button (B), but do not click OK when finished.

Gray 70%

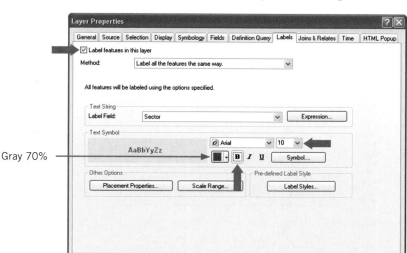

4   Click the Placement Properties button. Then click the Placement tab and select the "Only place label inside polygon" check box. Click the "Place one label per feature part" option, but do not click OK. Making these selections nicely fine-tunes the placement of labels.

5   Click the Conflict Detection tab and select High for Label Weight. Click OK. The selection of High for label weight means that the labels for other features—for example, the labels for streets when you create street labels—will not interfere with the sector label.

6   Click Symbol > Edit Symbol, and then click the Mask tab. In the Style area, click Halo. Then click OK three times. The halo option places a white halo around the sector numbers to make them stand out.

7   Double-click Streets and make label selections as shown in the figure.

Gray 50%

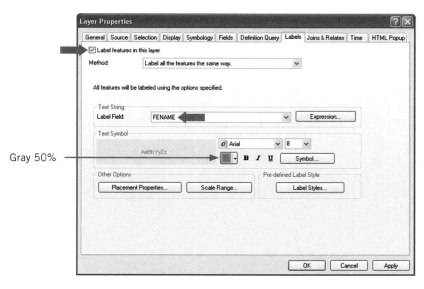

8   Click Placement Properties, click the Placement tab, and then click Curved. Select the "On the line" check box and clear the Above check box.

9 Click the Conflict Detection tab and select Low for Label Weight. Type **1** for the buffer. Click OK twice. The buffer results in fewer streets being labeled because field officers need relatively few streets labeled to provide context.

Set visible range for streets

When the map is zoomed so you can see all or most of Pittsburgh, the Streets layer becomes too cluttered, and thus should not be displayed at that scale. Streets are helpful when zoomed in close enough, starting at about the scale you are now with Pittsburgh's North Side filling up your map. Now, you will set a visible scale at which the Streets layer becomes visible. Note that if you decide to change a visible scale, the Visible Scale Range function you are about to use has an option to Clear Scale Range and start over again.

1 In the table of contents, right-click Streets and click Visible Scale Range > Set Minimum Scale. The streets disappear. If you were to zoom in one more foot, Streets would appear. Note the map scale field on the Standard toolbar. It has a value such as 1:46,211, depending on your screen size. In this case, 1 inch on the screen represents 46,211 inches on the ground. When you zoom in and features get bigger, the scale ratio is getting larger. For example, 1:1 is life size and 1:24,000 is large scale, but decreasing the scale ratio to 1:500,000 is small scale. So, when you set the minimum scale for visibility, you must use a larger scale or zoom in closer for the layer to be visible.

> 1:46,211

2 In the Map Scale field on the Standard toolbar, change the denominator so it is one unit smaller (for example, from 46,211 to **46,210** in this case, but, of course, your value will be different) and press ENTER. Streets appear.

3 On the Tools toolbar, click Full Extent. Streets disappear.

4 On the Tools toolbar, click Go Back To Previous Extent. Streets appear.

Add crime layer

Now that you have your basemaps in place, you need to add a crime layer to see the crimes that have occurred in the area. Even though the master file you use in this exercise has data for all of 2008, suppose that today is July 11, 2008, and you want to map the first seven serious crimes (FBI hierarchy values 1–7) for the last four weeks (28 days). The last seven days of the 28-day period ending July 11 (July 5–11) will have large bright symbols by crime type and the first three weeks (June 14 through July 4) will have faded, smaller versions of the same symbols. The approach to use is to add two copies of the offenses master file data to the map and add a definition query criterion to each copy to restrict the display to only the desired crime types and date range.

1 On the Standard toolbar, click Add Data. Then click Browse, go to the Police geodatabase, click Offenses2008, and click Add.

2 Right-click Offenses2008 in the table of contents and click Copy.

3 On the Menu bar, click Edit > Paste.

4 In the table of contents, click the text label of the top copy of Offenses2008, wait a moment, click it again to place the label in Edit mode, and rename the label by typing **Serious Crimes, 1st 3 Weeks**.

5   Do the same with the second copy of Offenses2008, but rename it **Serious Crimes, 4th Week**. Then turn this layer off.

6   Double-click the Serious Crimes, 1st 3 Weeks label, click the Definition Query tab, and click the Query Builder button. The query is as follows:

```
"DateOccur" >= date '2008-06-14' AND
"DateOccur" <= date '2008-07-04' AND
"Hierarchy" >= 1 AND
"Hierarchy" <= 7
```

You can try to build this query on your own, or follow the instructions in step 7. **Hint:** The result is easier to read if you add spaces to place each part of the query on a separate line as shown in the code.

7   Double-click "DateOccur" and click the >= button. Click the Get Unique Values button, scroll down to find the date '2008-06-14', and double-click it. Click the And button, double-click "DateOccur", and click the <= button. Scroll down to find the date '2008-07-04' and double-click it. Click And, double-click Hierarchy, and click >=. Type a space and **1**. Click And, double-click Hierarchy, and click <=. Type a space and **7**. Click Verify. If you have correct syntax that doesn't violate any rules, you'll get a message saying the expression was successfully verified and you can click OK. If you get an error message showing an invalid SQL statement, compare your query expression with the one shown, make corrections, and try verifying your syntax again.

8   After your query is verified, click OK twice.

9   In the table of contents, right-click Serious Crimes, 1st 3 Weeks, select Open Attribute Table, and verify that the date range and remaining crimes satisfy the query definition and that you have 1,211 records.

YOUR TURN   **YOUR TURN**   YOUR TURN   YOUR TURN   YOUR

Turn Serious Crimes, 1st 3 weeks off and turn Serious Crimes, 4th Week on. Then build a definition query that is similar to the one you built but for July 5–11, 2008. **Hint:** Open the definition query you just built, copy it by selecting it and pressing CTRL+C, and then paste it at the start of your new definition query by pressing CTRL+V. Edit the expression directly by typing to change dates as needed. You should have 416 records for the most recent week.

Symbolize crimes using unique values

You can use either plain, solid shapes for point markers, such as circles, squares, and triangles, or special point markers that are designed for crime mapping. Plain markers are always a good alternative, but in this exercise, use the special crime symbols provided by Esri. For many applications, it is important to use both shape and color to distinguish categories such as crime type. Shape by itself functions well and is sufficient for users who are color-blind and for map copies in black-and-white. Color helps the great majority of users who can see color by further distinguishing patterns. The rule is to use different, bright colors for different categories.

1   In the table of contents, turn the Serious Crimes, 4th Week layer on.

2   Double-click Serious Crimes, 4th Week and click the Symbology tab. In the Show panel under Categories, click "Unique values" and select Hierarchy for the Value field. Click Add All Values, and clear the "all other values" check box under Symbol, but do not click OK.

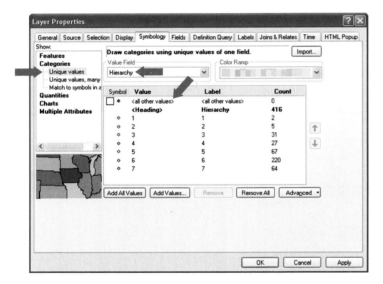

In the next step, you'll type labels for the hierarchy crimes by using the code table Hierarchy.xls, shown in the figure and available in the CodeTables folder in Data. Later, you use this table directly in ArcMap, but for now, you just need to see its contents for reference. Note that the values in this table are the Pittsburgh Police Bureau's code values and that there are some differences from the FBI's official code values.

3   In the Label column under Hierarchy, click the "1" and replace it by typing **Murder-Manslaughter**.

	A	B	C
1	Crime	Hierarchy	Part
2	Murder-Manslaughter	1	1
3	Forcible Rape	2	1
4	Robbery	3	1
5	Aggravated Assault	4	1
6	Burglary	5	1
7	Larceny - Theft	6	1
8	Vehicle Theft	7	1
9	Arson	8	1
10	Forgery	9	2
11	Simple Assault	10	2
12	Fraud	11	2
13	Embezzlement	12	2
14	Stolen Property	13	2
15	Vandalism	14	2
16	Weapons Violations	15	2
17	Prostitution	16	2
18	Sex Offenses	17	2
19	Drug Violations	18	2
20	Gambling	19	2
21	Family Violence	20	2
22	Drunken Driving	21	2
23	Liquor Law Violations	22	2
24	Public Drunkenness	23	2
25	Disorderly Conduct	24	2
26	Vagrancy	25	2
27	Other	26	2

4   Repeat step 3 for hierarchy labels 2–7 by
typing respectively: **Forcible Rape, Robbery,
Aggravated Assault, Burglary, Larceny-Theft,
Vehicle Theft**. Do not click OK.

Symbol	Value	Label	Count
☑ ◆	\<all other values>	\<all other values>	0
	\<Heading>	**Hierarchy**	**416**
◇	1	Murder-Manslaughter	2
◇	2	Forcible Rape	5
◇	3	Robbery	31
◇	4	Aggravated Assault	27
◇	5	Burglary	67
◇	6	Larceny-Theft	220
◇	7	Vehicle Theft	64

Use crime point marker symbols

1   In the Symbol column, double-click the small,
round point marker for Murder-Manslaughter,
and in the Search box at the top of the Symbol
Selector window, type **Crime Analysis** and press
ENTER. This search provides access to the
crime analysis symbols.

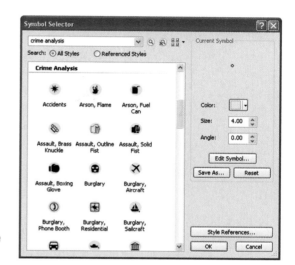

2   Scroll down the list of available point markers
until you find Homicide Evidence 🐾. Click it,
and then click Edit Symbol.

3   In the Symbol Property Editor, click the yellow
circle in the Layers area on the lower left,
change its color to Mars Red (the third row in the
red column), and click OK.

4   In the Symbol Selector window, change the size to 14 and click OK twice.

5   Repeat steps 1–4 with the following choices, all in size 14 and using color selections from the third row
of the color array:

 •   Forcible Rape: Rape point marker 🐏 , Electron Gold circle color.

 •   Robbery: Burglary point marker (there is no Robbery point marker, so you can use this one) 😵 ,
 Solar Yellow circle color.

 •   Aggravated Assault: Assault, Solid Fist point marker 👊 , Amethyst circle color.

 •   Burglary: Burglary, Residential 🏠 point marker, Big Sky Blue circle color.

 •   Larceny-Theft: Grand Theft point marker 💰 , Medium Apple circle color.

 •   Vehicle Theft: Grand Theft Auto point marker 🚗 , Tourmaline Green circle color.

6   Click OK.

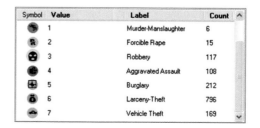

Symbol	Value	Label	Count
	1	Murder-Manslaughter	6
	2	Forcible Rape	15
	3	Robbery	117
	4	Aggravated Assault	108
	5	Burglary	212
	6	Larceny-Theft	796
	7	Vehicle Theft	169

7   **Zoom in on Police Sector 1-5 and include at least half the area of surrounding police sectors.**

8   **Save your map document.**

Turn on and symbolize Serious Crimes, 1st 3 Weeks using the same symbols, only smaller (size 10) and the same hue but lighter (first row of color palette). **Hint:** You can take a shortcut by importing symbolization from Serious Crimes, 4th Week, and then modify the symbols. Click the Import button under the Symbology tab of the Layer Properties window for Serious Crimes, 1st 3 Weeks and complete the steps. Notice several persisting hot spots for larcenies and robberies and a possible emerging hot spot for vehicle thefts in the resulting map—all highlighted by circles (see figure map). The robbery hot spot overlaps Police Sectors 1-4 and 1-5, and thus requires some coordination among the field officers for each sector. Can you find any more potential hot spots affecting multiple police sectors? You can draw circles around hot spots by going to Customize > Toolbars > Draw, clicking the Circle tool, and then double-clicking circles to change the outline color and remove the color fill. When you are finished, save your map document.

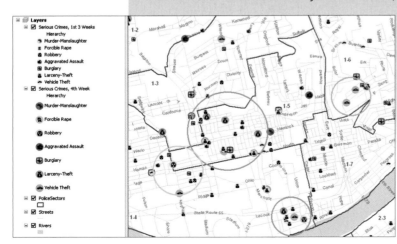

The week ending July 11, 2008, was a hard one for Police Sector 1-5 and for victims, with a homicide, a robbery, miscellaneous aggravated assaults, burglaries, larcenies, and a vehicle theft all occurring in the sector.

Set field properties

The results table for the Identify tool and the attribute table for crimes have some extraneous columns that were created by ArcGIS for the two crime layers for internal processing but are not needed by the crime analyst. Other columns have cryptic names chosen by the designer of the attribute table. You can help the user by not displaying the fields that are not being used and renaming others that are being used.

1　Right-click Serious Crimes, 1st 3 Weeks; select Properties; and click the Fields tab. Click to clear all fields above CCN.

2　Also, click to clear TimeOccNum, which is used only for data processing.

3　Click the DateOccur field name (leaving the field turned on) and type a space in the alias name DateOccur so it reads **Date Occur**.

4　Create a few more aliases for practice.

5　Click OK.

6　Open the attribute table for the Serious Crimes, 1st 3 Weeks layer to see the results.

7　Close the table.

CCN	Address	Date Occur	Week Day	Time Occur
2008017966	5102 DRESDEN WY	6/10/2008	3	12:40:00 AM
2008017977		6/10/2008	3	3:20:00 AM
2008017981	703 LARIMAR AV	6/10/2008	3	12:30:00 AM
2008018015	5536 JACKSON AV	6/10/2008	3	12:30:00 PM
2008018049	120 5TH AV	6/10/2008	3	3:00:00 PM
2008018056	7321 TIOGA ST	6/10/2008	3	1:00:00 PM

YOUR TURN　YOUR TURN　YOUR TURN　YOUR TURN　YOUR

Repeat steps 1–6 for Serious Crimes, 4th Week.

Set selectable layers

1　At the top of the table of contents, click List By Selection.

2　Click the toggle keys for Police Sectors, Rivers, and Streets so these layers are not selectable.

3　Click List By Drawing Order.

4　Try the Identify tool on a few crime points to see that ArcMap displays only the fields set to be visible.

Add hyperlinks to crime points

Suppose that the police organization's policy is to add the offense report narrative statement and other materials as hyperlinks to all murder-manslaughter offenses. You can link one or more documents or files to any mapped feature, and then view them by clicking the feature on the map.

1　Click the Identify tool.

2　Click the center of the Murder-Manslaughter point marker in Police Sector 1-5 and choose "Selectable layers" in the "Identify from" field.

3　In the top panel of the Identify tool, right-click the value under Serious Crimes, 4th Week (i.e., Christian) and click Add Hyperlink. Click Browse, go to Data > RawData, and double-click Narrative2008021985.docx. Click Open > OK.

4　Close the Identify window.

5　On the Tools toolbar, click the Hyperlink tool. Notice that the Murder-Manslaughter point marker in Police Sector 1-5 now has a blue dot in the center, indicating that it has hyperlinks. With the bottom tip of the Hyperlink pointer, click the blue dot. The narrative document for the murder opens in Microsoft Word. After reading the narrative, close it.

N YOUR TURN **YOUR TURN** YOUR TURN YOUR TURN YOU

Add a second hyperlink to the Murder-Manslaughter point in Police Sector 1-5 that goes to the image Gun20080221985.jpg in the RawData folder. Try using the Hyperlink tool on the point. View the image, and then close it.

Save layer files

You can save the work you did in symbolizing crimes for use in other maps. As discussed at the start of this chapter, the saved work is called a "layer file," and you can import it and apply it to a new crime map instead of having to once again carry out all the interactive work every time you create a map.

1　In the table of contents, right-click Serious Crimes, 1st 3 Weeks, select Save As Layer File, and save to your chapter 4 folder in MyExercises.

2　Repeat step 1 for Serious Crimes, 4th Week.

In the next step, you'll try out the Serious Crimes, 1st 3 Weeks layer by adding Offenses2008 to your map document and symbolizing it with the layer file. Imagine that you do not have any other layers in the map document from which you could import the symbols (although, of course, you have Serious Crimes, 1st 3 Weeks, which you could use as an alternate map layer).

3　On the Standard toolbar, click Add Data. Then click Browse, go to the Police geodatabase, and click Offenses2008. Then click Add.

4　In the table of contents, double-click Offenses2008, click the Symbology tab, and click the Import button. Go to your chapter 4 folder in MyExercises and double-click the Serious Crimes, 1st 3 Weeks layer. Click OK three times. Now that you can see how layer files work, you can recycle symbolization and remove Offenses2008 from your map document.

5　In the table of contents, right-click Offenses2008 and click Remove.

6　Save your map document, but do not close it.

Tutorial 4-2

Building an early-warning system for investigators

You start this tutorial by recycling your pin map from tutorial 4-1 as a starting map document for the burglary early-warning map, keeping just the contextual map layers. Then you build two burglary choropleth maps and finally the burglary pin map that allows you to drill down within the layers to see greater detail, ending up with all three maps in the same map document. You use threshold layers to turn map layers on and off, depending on how closely you are zoomed in, which makes this map document dynamic for the user. When zoomed out, you see a choropleth map; when zoomed in, you see the pin map and streets.

Reuse previous map document

To get started, you can reuse the map document from tutorial 4-1. If you did not create the pin map, one option is to do the steps in the previous sections titled "Create new map document," "Add layers to map document," and "Set visible range for streets." Otherwise, you can use Tutorial4-1.mxd from the chapter 4 folder in FinishedExercises.

1 If you closed ArcMap, open it, and then openTutorial4-1.mxd.

2 Save your map document as **Tutorial4-2.mxd** to your chapter 4 folder in MyExercises.

3 Remove both crime map layers. Also, remove any drawn circles (by clicking each one with the Select Elements tool on the Standard toolbar and pressing DELETE.) Zoom to full extent.

In the next step, you'll reduce the size and prominence of police sector labels, because the map user will mostly be viewing police sectors for the entire city. These labels needed to be much larger when zoomed in on a particular police sector for the officers' pin maps, but they need to be smaller now for overall viewing.

4 In the table of contents, double-click the Police Sectors label and click the Labels tab. Change the text symbol to size 8 and turn Bold off.

5 Click Symbol > Edit Symbol and click the Mask tab. Reduce the halo size to 1. Click OK three times.

6 Save your map document.

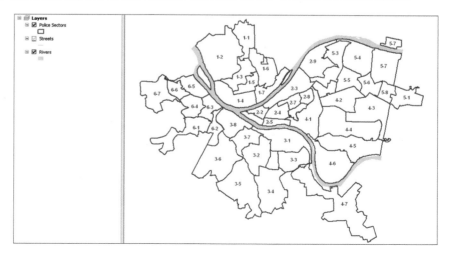

Join data for a choropleth map

1   On the Standard toolbar, click Add Data. Then click the Browse button, go to the Pittsburgh geodatabase, and add another copy of PoliceSectors to the map document.

2   Change the name of this map layer from PoliceSectors to **Burglary Level: July 2008**.

3   Again, click Add Data. Then click Browse, go to the chapter 9 folder in FinishedExercises, and add BurglarySeriesComplete200810$ from the BurglarySeriesComplete200810 spreadsheet to the map document. The added table has the number of burglaries per month by police sector for January through October 2008. In the next step, you'll join this table to the Burglary Level: July 2008 polygon layer, and then restrict the data to July 2008 by a definition query. Assume that it is the end of July (and ignore the data for August through October).

Sector	Year	Month	Burglary
1-1	2008	1	2
1-1	2008	2	1
1-1	2008	3	0
1-1	2008	4	5
1-1	2008	5	6
1-1	2008	6	4
1-1	2008	7	1
1-1	2008	8	2
1-1	2008	9	4
1-1	2008	10	0
1-2	2008	1	10

4   In the table of contents, right-click Burglary Level: July 2008, click Joins and Relates > Join, and make the selections shown in the figure.

5   Click OK. Nothing appears to change, but now the table for Burglary Level: July 2008 has additional fields (Year, Month, and Burglary) joined to the right side of the table. There were 10 possibilities, January through October data, for each police sector, so ArcMap arbitrarily chose month 1. Take a look. Next, you'll force ArcMap to use month 7, July.

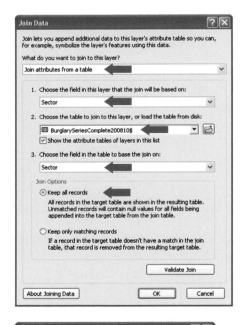

6   In the table of contents, double-click Burglary Level: July 2008, click the Definition Query tab, and click the Query Builder button. Use the query builder to build the query as shown in the figure. (Type a space and **7** for the month rather than using the Get Unique Values button.)

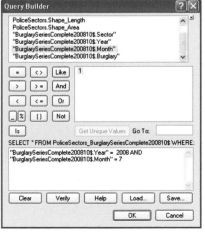

7 Click OK > OK > Open to open the Burglary Level: July 2008 attribute table. Verify that each police sector has burglary data for July 2008.

OBJECTID *	Shape *	Sector	Shape_Length	Shape_Area	Sector	Year	Month	Burglary
1	Polygon	3-4	37667.159272	55888083.839381	3-4	2008	7	9
2	Polygon	4-7	87456.366401	98370955.580853	4-7	2008	7	2
3	Polygon	3-5	45996.314009	72045057.511819	3-5	2008	7	5
4	Polygon	3-3	23594.776021	28347798.547509	3-3	2008	7	4
5	Polygon	6-2	29.472495	0.236088	6-2	2008	7	3
6	Polygon	3-8	91.309471	21.63675	3-8	2008	7	1

8 Close the table and save your map document.

Create custom scale to symbolize a choropleth map

Now, you need to design a numerical scale similar to that of a bar chart to yield discrete classes. For example, if you wanted to create a bar chart of the burglary level by police sector for July 2008, you might choose equal-interval classes of 1–5, 6–10, and so on for the number of burglaries in each sector. Choropleth maps need the same kind of numerical scale and classes so that you can assign a color to each interval, with increasingly darker shades from the same color hue representing intervals with increasing levels of crime.

1 Double-click the Burglary Level: July 2008 layer and click the Symbology tab. In the Show panel, click Quantities > Graduated colors. In the Value field, select Burglary and click the Classify button.

2 In the Classification panel, select Manual for Method and 7 for Classes. Crime distributions are skewed to the right, which means they have many areas with relatively low or medium levels of crime and a few areas with extremely high levels of crime. Such a distribution is best portrayed by using intervals of increasing width. Here, you can use powers of 2 for the end points (also known as break points) of the intervals: 0, 1–2, 3–4, 5–8, 9–16, 17–32, and 33 and higher. This particular scale will work well with many types of crime at the police sector level, using monthly data.

3 In the Break Values panel, click 20 and type 999 (this is a "trick" representing a very large number of crimes thought never to be exceeded in a sector), click 14 and type 32, and continue as follows:

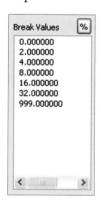

Break Values %
```
0.000000
2.000000
4.000000
8.000000
16.000000
32.000000
999.000000
```

- 11 becomes **16**

- 8 remains **8**

- 5 becomes **4**

- 3 becomes **2**

- 1 becomes **0**

4 Click OK.

5 If the color ramp is not yellow to orange to brown, make that selection.

6 Double-click the yellow color chip for the 0 range and change its color to Arctic White (not Hollow). Click OK.

7 Click the Label heading above the values at the right, click Format Labels, and change the number of decimal places from 6 to 0. Click OK.

8 Click the cell with "Label 33 - 999" and change the text to **33 and higher**.

9 Click OK.

10 In the table of contents, turn on and then right-click Burglary Level: July 2008. Click Save As Layer File and save to your chapter 4 folder in MyExercises. This completes the map for burglary levels in July 2008. Police Sectors 3-1, 3-2, 5-3, and 5-7 are all in the low end of the 17–32 burglaries class, with Police Sector 3-2 having the highest number of burglaries, at 20. (You can confirm this with the Identify tool.)

Create "Change map" data using spreadsheet software (optional exercise)

At this stage, you need a new data attribute for changes in the number of burglaries, so this section has instructions for creating that attribute by using Microsoft Office Excel. You can skip this section, however, and get the needed result from the file in the chapter 4 folder in FinishedExercises. To calculate the difference between July burglaries and June burglaries in 2008 by police sector, you need to have July burglaries and June burglaries on the same line in a table. The June burglaries, in this case, are known as "lagged" data because they lag, or are behind, July burglaries on the calendar. Not many software packages have the ability to create lagged data, except statistical packages where lags are common. So next, you'll use Excel to copy and paste monthly data to create the needed lag.

1 Start Microsoft Office Excel 2007, click the Microsoft Office button 🔳, and click Open. Go to the chapter 9 folder in FinishedExercises and open the BurglarySeriesComplete200810 spreadsheet.

2 Click Office > Save As. Type **BurglarySeriesCompleteChange200810.xls** for File Name and save to your chapter 4 folder in MyExercises. The data is already sorted by sector, then by year, and then by month, which is necessary for steps 3–6 to create the lag.

3 Click in cell E1 and type **Lag**. Click in cell D2, press SHIFT, scroll down to the bottom of the table, and click cell D461 to select that range of cells. Then press CTRL+C to copy the data.

	A	B	C	D	E
1	Sector	Year	Month	Burglary	Lag
2	1-1	2008	1	2	
3	1-1	2008	2	1	2
4	1-1	2008	3	0	1
5	1-1	2008	4	5	0
6	1-1	2008	5	6	5
7	1-1	2008	6	4	6

4 Scroll to the top of the sheet, click in cell E3, and press CTRL+V to paste the selected data.

5 In cell F1, type **Change**, and then in cell F3, type **=D3-E3**.

	A	B	C	D	E	F
1	Sector	Year	Month	Burglary	Lag	
2	1-1	2008	1	2		
3	1-1	2008	2	1	2	=D3-E3
4	1-1	2008	3	0	1	

6 Press ENTER. Click in cell F3 and press CTRL+C. Then click in cell F4, press SHIFT, scroll down to cell F461, and press CTRL+V.

	A	B	C	D	E	F
1	Sector	Year	Month	Burglary	Lag	Change
2	1-1	2008	1	2		
3	1-1	2008	2	1	2	-1
4	1-1	2008	3	0	1	-1
5	1-1	2008	4	5	0	5
6	1-1	2008	5	6	5	1
7	1-1	2008	6	4	6	-2

7 Click the F column heading to select the entire column, right-click in the F column, and click Copy. Then right-click in the F column and click Paste Special. Click Values and click OK. The paste special step replaced stored expressions with values, making the file more flexible for future use.

Delete initial lags, using spreadsheet software (optional exercise)

Now, you need to delete change values for all rows that have Month = 1, where there is no legitimate lag value for a sector.

1 Press SHIFT and drag the mouse pointer across the column headings for columns A through F to select those columns.

2 On the Excel Menu bar, click Data > Sort and select Month. Click OK.

3 Click in cell F2 (which is blank), and then press and drag the lower-right corner of the cell down through all the month 1 rows, through row 47, and release. This step copies the blank row from F2 throughout the rows in which month = 1.

4 Click the column A heading, press SHIFT, and drag the pointer across the headings through column F to select those columns.

5 On the Excel Menu bar, click Data > Sort.

6 Click Add Level twice so there is a total of three Sort By rows.

7 Make selections as shown in the figure.

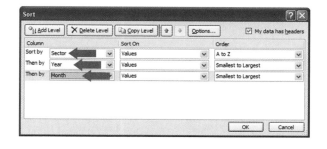

8   **Click OK twice.** The lag data is now finished. There can be no lag value for the oldest data point in a series, so each sector's month 1 has a blank (null value).

9   **Save your file and exit Excel.**

	A	B	C	D	E	F
1	Sector	Year	Month	Burglary	Lag	Change
2	1-1	2008	1	2		
3	1-1	2008	2	1	2	-1
4	1-1	2008	3	1	1	0
5	1-1	2008	4	5	1	4
6	1-1	2008	5	6	5	1
7	1-1	2008	6	4	6	-2
8	1-1	2008	7	1	4	-3
9	1-1	2008	8	2	1	1
10	1-1	2008	9	4	2	2
11	1-1	2008	10	0	4	-4
12	1-2	2008	1	10	0	
13	1-2	2008	2	11	10	1
14	1-2	2008	3	14	11	3

YOUR TURN   YOUR TURN   **YOUR TURN**   YOUR TURN   YOUR TURN   YOU

Add another copy of PoliceSectors from the Pittsburgh geodatabase to Tutorial4-2.mxd. Then do the following steps:

- In the table of contents, change its name to **Burglary Changes: July-June 2008** and turn it off for now.

- Add BurglarySeriesCompleteChange200810.xls to your map document. If you did not create this file in the preceding optional set of steps, use the copy in the chapter 4 folder of FinishedExercises.

- Join BurglarySeriesCompleteChange200810.xls to Burglary Changes: July-June 2008.

- Build a definition query for Burglary Changes: July-June 2008 as follows:

  "BurglarySeriesCompleteChange200$.Year" = 2008 AND
  "BurglarySeriesCompleteChange200$.Month" = 7

Be sure to type the space and **7**. Then check this map layer's attribute table to make sure the definition query and join work.

Turn off Burglary Level: July 2008 in the table of contents. Turn on Burglary Changes: July-June 2008 and use it to build a choropleth map for burglary changes, using the Change attribute joined to that layer's attribute table. The map should look like the figure map when finished.

- Start with seven natural break classes for the Change attribute. Then use the Manual classification method with the following break points: -13, -7, -3, 2, 6, 12, and 999 (entered from right to left in this list and from the bottom up in the Break Values panel). Notice that the result is a symmetrical scale with the negative intervals mirroring the positive intervals in increasing widths but that the specification of negative break points is tricky and different from that of the positive ones.

- Double-click individual color chips and select blue and red colors so that the darkness increases with absolute magnitude and leaves white in the middle.

(continued)

- Edit the labels of categories as shown in the figure map below.

- In the table of contents, move the Police Sectors layer to the top (click List By Drawing Order first).

- Right-click Burglary Changes: July-June 2008, select Save As Layer File, and save to your chapter 4 folder in MyExercises.

- Save your map document.

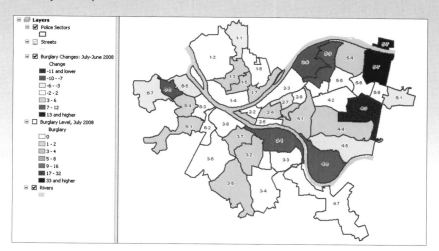

There is a lot of information from the pair of choropleth maps you built: for example, Police Sectors 5-7, 5-3, 4-6, and 3-1 are newly hot in July, while sectors 1-2 and 4-2 have persisted as moderately hot in June and July. Next, you'll add layers for drilling down to points. You will add July 2008 burglaries, disorderly conduct, and vagrancy crimes as point layers. Disorderly conduct and vagrancy crimes tend to be leading indicators of burglaries—that is, they tend to occur first and lead to burglaries—so adding these layers provides additional diagnostic information.

Create size-graduated point markers

It is important to be able to identify burglary locations that have repeat burglaries within the same time period. As research by K. J. Bowers et al. (2004) has shown, such locations and vicinities are likely to have future burglaries. ArcToolbox has a tool that tabulates the number of records at the same address within a given time period. You use that tool in this exercise. Note that normally you will be viewing the new layer zoomed in on a portion of the jurisdiction for drilling down to problem police sectors, so you need to choose relatively large point markers for that.

1   In the table of contents, turn the two choropleth maps off and add a copy of Offenses2008 from the Police geodatabase to your map document.

2   Add the following definition query to this map layer:

   "DateOccur" >= date '2008-07-01' AND

   "DateOccur" <= date '2008-07-31' AND

   "Hierarchy" = 5

3   On the Standard toolbar, click the ArcToolbox button .

4   In the ArcToolbox window, expand the Spatial Statistics toolbox and then the Utilities toolset. Then double-click the Collect Events tool.

5   For Input Incident Features, select Offenses2008.

6   For Output Weighted Point Feature Class, click Browse, go to the chapter 4 geodatabase in MyExercises, and type **BurglaryJuly2008Graduated** for Name. Click Save > OK. Wait for ArcMap to finish processing. (See the Collect Events window for progress information.) Ignore any warning messages. The result is size-graduated point markers where the size increases along with the number of burglaries at the same address in July 2008, but the symbols need to be adjusted. By opening the attribute table and sorting by the ICOUNT column in descending order, you can see that there are 10 burglary locations that had two burglaries each in that month, and the rest had one burglary.

7   Close the ArcToolbox window.

Symbolize size-graduated point markers

1   In the table of contents, double-click the BurglaryJuly2008Graduated layer, click the Symbology tab, and then click Quantities > Graduated Symbols. In the Value field, select ICOUNT. Click Classify and select Quantile for Classification method. Click OK.

2   Double-click the symbol for the 1 range and change its size to 8 and color to Medium Apple (seventh column, third row in the color chips table).

3   Do the same for the 2 range, except change its size to 12 and color to Leaf Green (seventh column, fifth row). Click OK in the Layer Properties window.

4   In the table of contents, move the Streets layer above the two choropleth map layers.

5   Turn Burglary Changes: July-June 2008 on and zoom in on the South Side of Pittsburgh as shown in the figure map. The original crime offense layer, Offenses2008, is still needed in the table of contents, even though it is under BurglaryJuly2008Graduated and not visible. The new layer, BurglaryJuly2008Graduated, has none of the attributes needed by field officers, but Offenses2008 does have these attributes.

6   In the table of contents, click List By Selection and make Offenses2008 the only selectable layer. Try using the Identify tool on burglary points with the "Identify from" field set to selectable layers. In particular, try the tool on points with two burglary incidents and note that you can access the records for both incidents. (Remember to click just to the right of one of the two horizontal, dashed lines in the top panel of the Identify results to select a record.)

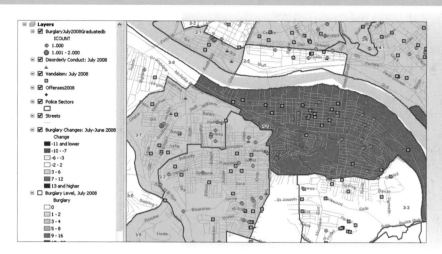

Add two more copies of Offenses2008 from the Police geodatabase to your map document. Set the definition query for one copy to the July 2008 date range and hierarchy 14 (vandalism) and the other copy to the same date range and hierarchy 24 (disorderly conduct). **Hint:** Copy and paste the definition query from your original copy of Offenses2008 and modify the hierarchy value for each new layer. Symbolize these layers to your liking and use single point markers (not graduated point markers).

Set visible scale ranges

The final step in completing the early-warning map is to set visible scale ranges. You need the choropleth maps to turn on when zoomed out far enough and the Streets and point crime layers to turn off. The opposite is needed when the map is zoomed in close enough.

1   Turn all layers on and zoom in on the North Side of Pittsburgh, the area totally above rivers.

2   Right-click Streets and click Visible Scale Range > Set Minimum Scale.

3   Right-click each of the four crime point layers, one after the other, and click Visible Scale Range > Set Minimum Scale. The check box for turning the layer on dims, indicating that this layer is now out of visible range.

4   Right-click each of the choropleth maps, one after the other, and click Visible Scale Range > Set Maximum Scale.

5   Zoom to full extent and try zooming in and out of various police sectors. The drilldown capability demonstrated here is one of the best tools of crime analysis.

6   Save your map document.

Tutorial 4-3

Building a map for public use

Maps for the public, like the one you work on in this tutorial, need to be self-documenting and easy to interpret. A map layout, such as the one in chapter 2, is ideal for this purpose. It has components, such as a title, legend, and scale bar, that help with the interpretation and use of a map. The map document you use in this tutorial has a new basemap layer, Pittsburgh neighborhoods. Although police are concerned with police sectors, the public is more concerned with neighborhoods.

Open map document

To keep the focus on building a map layout, Tutorial4-3.mxd has a finished map document, except for the layout. It includes property crimes and related offenses for July 2008 symbolized with unique point markers. The crime layer has offense locations moved from street addresses to random points within the same block. The map is zoomed in on the Shadyside neighborhood, where many university students live off campus. Suppose you are a new student at the University of Pittsburgh, Chatham College, Carlow College, or Carnegie Mellon University. You want to live in Shadyside; and so you want information on property crimes there because you're worried whether your car, stereo, or computer might be stolen.

1 **Open Tutorial4-3.mxd in ArcMap.** Notice that the unique symbols for the selected crimes have color coding roughly following the color spectrum, from hot to cold, arranged by seriousness according to the FBI hierarchy, where burglary is the most serious crime. As you can see, Shadyside has its share of larcenies, vehicle thefts, and vandalism offenses.

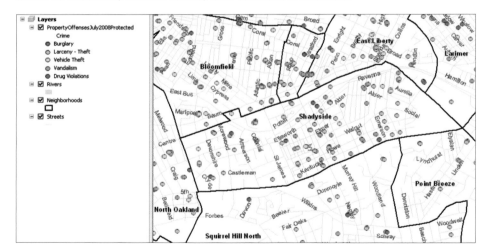

2 **Save your map document to your chapter 4 folder in MyExercises.**

Change layout page orientation

ArcMap starts with a bare-bones layout in Portrait Page orientation. You can change the orientation to Landscape, which offers a better size of map, and then add several layout components.

1   Click View > Layout View.

2   Right-click in any white area of the main panel that has the layout, and select Page and Print Setup. In both the Paper panel and Map Page Size panel, click Landscape. Click OK.

3   Create vertical guide lines for resizing and locating the map by clicking the horizontal ruler at the top of the window at 1 and 8.5 in. If you make a mistake, you can right-click the arrow at the top of a guide line and clear it. Alternately, you can drag the guide line to a new position.

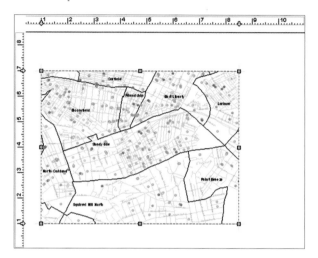

4   Similarly, for horizontal guide lines, click the vertical ruler at 1 and 7 in.

5   Click the map, so you can drag it by the corners, and then resize it to fit the rectangle formed by the four guide lines. (The map will snap to the guide lines.)

6   Right-click in the white area of the layout and select Zoom Whole Page.

Add title to a layout

1   From the Menu bar, click Insert > Title.

2   Move the title to center it above the map, and then double-click the title and change its text to **Property and Related Crimes: July 2008**. Click Change Symbol and select size 22, style B (bold) for the font. Click OK twice.

Add legend to a layout

1   Create a new guide line at 8.7 in on the horizontal ruler, to the right of the 8.5 in guide line.

2   From the Menu bar, click Insert > Legend.

3   Move all layers except PropertyOffensesJuly2008Protected from the right panel to the left panel by clicking each layer and the < button. Then click Next four times and click Finish.

4   Move the legend to the right of the map so its lower-left corner is on the intersection of the 1 in horizontal guide line and the 8.7 in vertical guide line. The legend extends too far to the right, but you can fix that problem by deleting its long title.

5   Right-click the legend and select Convert To Graphics. Right-click the legend again, select Ungroup, and click anywhere in a white area of the layout to clear legend elements. Click the text box for PropertyOffensesJuly2008Protected and press DELETE.

6   Drag the Legend text box down to just above the Crime text box.

7   Draw a rectangle around all remaining legend elements to select them, right-click the collection, and select Group.

8   Resize the legend to make it larger but still anchored to the guide lines. You may need to adjust the placement of the Legend and Crime text box elements to keep them above the other legend elements.

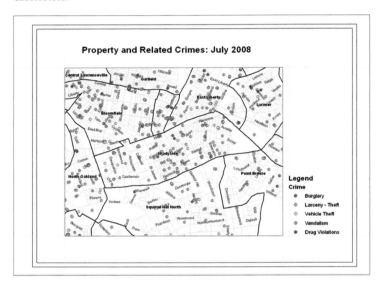

Add scale bar to a layout

1   From the Menu bar, click Insert > Scale Bar.

2   In the Scale Bar Selector, click the uppermost scale style, click the Properties button, and select Miles for Division Units. Click OK twice and move the scale below the map.

3   Click the right side of the scale and drag it by the corner to adjust the width so that the scale maximum is 1 mi.

Add note to a layout

1   From the Menu bar, click Insert > Text.

2   Move the text to the upper right side of the map under the intersection of the 7 in horizontal and 8.7 in vertical guides.

3   Double-click the text and type the following (with hard returns for new lines):

Note: The map plots crimes

randomly within blocks in

which they occurred to

protect privacy.

4   Click Change Symbol, select Arial size 11, and click OK. Click the Justify Left button 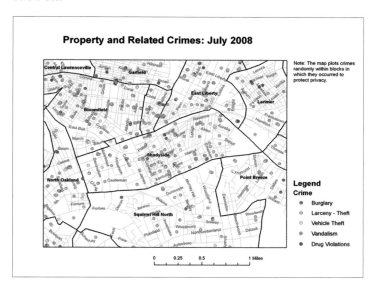. Click OK.

Add neatline to a layout

1   From the Menu bar, click Insert > Neatline.

2   In the Placement panel, click the "Place inside margins" option, choose None for Background, and click OK.

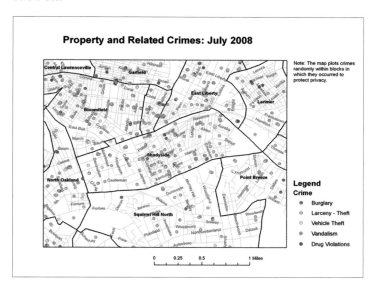

Export layout as an image file

Now, you can export your finished map as a high-quality image file for possible use in a Word document or PowerPoint presentation.

1   From the Menu bar, click File > Export Map.

2   Click Browse and go to your chapter 4 folder in MyExercises. Select JPEG for File Type, 300 dpi for Resolution, and type **Shadyside July 2008 Property Crimes.jpg** for File Name. Click Save.

3   Go to the saved file in a My Computer window, right-click the file, and select Preview to see how the map looks. It will look very good because the resolution is 300 dpi, which is publication quality.

4   Save your map document and close ArcMap.

Assignment 4-1

Build an auto theft pin map

Auto thieves, like burglars, are often creatures of habit. Some auto thieves steal cars for transportation, so they steal from locations where they live, where their friends live, or where they hang out. Professional car thieves also have favorite hunting grounds. As a result, crime maps are useful for targeting patrols to prevent auto thefts and apprehend the thieves.

For this assignment, build a sophisticated pin map for auto thefts.

Create new map document

Create a new map document called **Assignment4-1YourName.mxd** and save it to your assignment 4-1 folder in MyAssignments. Use relative paths for your map document. Add the following map layers and data:

+ Streets, Rivers, PoliceSectors, and ZoningCommercialBuffer from the Pittsburgh geodatabase in the Data folder.

+ OffenseMaster200811101, a feature class in the chapter 9 geodatabase in the FinishedExercises folder. Create an assignment 4-1 geodatabase in your assignment 4-1 folder. Then copy the OffenseMaster200811101 feature class to this geodatabase and use it from there.

+ Police Codes$ data sheet of the PittsburghPoliceOffenseCodes spreadsheet from the CodeTables folder.

Symbolize basemap layers

Symbolize the basemap layers using best practices. Include the following specific elements:

+ Add a visible scale range to Streets so they turn on when you are zoomed in on about a quarter of Pittsburgh or closer in. Label Streets and PoliceSectors.

+ Symbolize the ZoningCommercialBuffer shapefile with categories by using unique symbols, making the Commercial field value 0 hollow and value 1 cropland (search for Cropland in the Symbol Selector window). Change the colors for cropland to a background color, such as dark gray.

Prepare and symbolize offense data

Suppose the Auto Theft Squad wants the following offense codes included in its maps:

- 392110 MOTOR THFT/AUTO
- 392114 MTR THEFT ATTEMPT (AUTO-PASSGR)
- 392118 MOTOR THEFT (AUTO)-PLATE ONLY

The OffenseMaster200811101 feature class stores the crime codes in two fields, UCR and OffenseCode. For example, instead of having 392110 in one field, it has UCR = 3921 and OffenseCode = 10.

Combine these two fields into a new field, PoliceCode, that makes your definition query simpler by using the following steps:

1 Open the appropriate attribute table, click the Table Options button ⊟▾, and select Add Field.

2 Call the new field **PoliceCode** and set Type to Text with length **8**.

3 Right-click the new column heading, PoliceCode, click Field Calculator, and create the expression **PoliceCode = [UCR] & [OffenseCode]**. Click OK.

You will have values such as 392114 for PoliceCode. If you make a mistake, you can recalculate the same field.

Suppose the Auto Theft Squad wants 28-day, moving-data-window pin maps with detailed auto theft offense codes. It wants the most recent week's offenses to have more prominent point markers to provide a temporal context. Suppose it is September 30, 2008, so the time periods are 9/3/2008 through 9/23/2008 and 9/24/2008 through 9/30/2008.

Add a definition query to OffenseMaster200811101 that limits the data to the first three weeks of the data window and the desired crime codes.

Hint: You will need both AND and OR connectors. Put parentheses around the date conditions and around the PoliceCode conditions so that you have two sets of them:

        (date conditions with AND) AND (PoliceCode conditions with OR)

Be sure the OffenseMaster200811101 attribute table has the correct query logic.

Copy and paste OffenseMaster200811101 to the table of contents and create a similar definition query for the last week of the data window.

Symbolize the offense layers by using an approach similar to the one for the uniformed officers' pin map created earlier in this chapter. Use plain symbols that have outlines such as circles, squares, and triangles. Change the labels in the table of contents to make them descriptive of the codes.

Conduct a hot spot analysis for auto thefts (any of the three codes, 392110, 392114, 392118) on the North Side. Use the following criteria:

+ Make a layout with a map, title, legend, and scale bar. Modify the map extent and layout title for each required map listed as follows.

+ Prepare a Word document called **Assignment4-1YourName.docx** and save it to your assignment 4-1 folder with the following elements:

    ◊ Title, your name, and date

    ◊ Hot spot criteria listed above

    ◊ **Map4-1aYourName.jpg** image zoomed in on the North Side of Pittsburgh (north of the Allegheny and Ohio rivers)

    ◊ Map images—**Map4-1bYourName.jpg** and **Map4-1cYourName.jpg**—and a paragraph analyzing the two auto theft hot spots (persisting or emerging), zoomed in, that are best for targeted patrol in the area. Draw circles around the hot spots.

What to turn in

Use a compression program to compress and save your assignment 4-1 folder to **Assignment4-1YourName.zip**. Turn in your compressed file.

Assignment 4-2

Build auto squad choropleth maps

For this assignment, build the wide-area scan maps for auto thefts—one choropleth map for crime level and one choropleth map for the change in crime levels. You can work this assignment by itself or together with assignment 4-1. If you do both problems, combine solutions to yield a complete early-warning system. If you have not already done assignment 4-1, read the first paragraph in that assignment on the nature of auto thieves.

Create new map document

If you have finished assignment 4-1, open that map document, save it as **Assignment4-2YourName.mxd** to your assignment 4-2 folder in MyAssignments, and use it as your starting point.

Otherwise, start by creating a new map document called **Assignment4-2YourName.mxd** and save it to your assignment 4-2 folder. Use relative paths for your map document. Add the following map layers and data:

+ Rivers and PoliceSectors from the Pittsburgh geodatabase in the Data folder.

+ AutoAgg09.csv from the chapter 4 folder in FinishedExercises. This text file includes Sector, Auto08 (number of auto thefts by sector between 8/6/2008 and 9/02/2008), Auto09 (number of auto thefts by sector between 9/3/2008 and 9/30/2008), and Diff (Auto09 − Auto08).

Symbolize map layers

Symbolize the basemap layers using best practices. Create two choropleth maps: one for auto theft levels using the Auto09 attribute and one for changes in auto thefts using the Diff attribute. Use the same scales as in tutorial 4-2.

Optional

If you have the pin maps from assignment 4-1 included in your map document, add visible scale ranges so the layers turn on and off as needed at the scale of about a quarter or less of the city area. Make a layout with a map, title, legend, and scale bar.

Prepare a Word document called **Assignment4-2YourName.docx** and save it to your assignment 4-2 folder with the following elements:

+ **Map4-2aYourName.jpg** image with full map extent showing auto theft levels and **Map4-2bYourName.jpg** image with full map extent showing auth theft changes

+ Paragraph analyzing the best sector to target patrols, based on crime level and change

What to turn in

Use a compression program to compress and save your assignment 4-2 folder to **Assignment4-2YourName.zip**. Turn in your compressed file.

References

Bowers, K. J., S. D. Johnson, and K. Pease. 2004. Prospective hot-spotting: The future of crime mapping? *British Journal of Criminology* 44 (5): 641–58.

OBJECTIVES

Create attribute queries
Create spatial queries

Chapter 5

Querying crime maps

Conducting attribute and spatial map queries is at the heart of crime analysis. In working with a crime mapping and analysis system that uses master map layers like the ones you create in part 3 of this book, you will be ready to query this data for unique day-to-day information as crime problems emerge. This chapter provides a structured set of query types that cover most unique, one-time crime information needs.

Crime query concepts

Information systems provide two kinds of information, periodic information that is regularly updated and one-time (or ad hoc) information that arises from recent events. Periodic information is anticipated and provided every day, every week, every month, or every other relevant time interval as needed. All the maps you use in chapter 3 and build in chapter 4 are periodic maps that are used and reused on a regular basis. One-time information that covers ad hoc situations, on the other hand, is not anticipated, but comes in response to unique crime patterns or as a result of particular police interventions. Examples of ad hoc scenarios include crime patterns involving serial crimes, a gang rivalry, a special event such as a rock concert, or a police crackdown on a drug hot spot or on crime-prone land uses such as bars.

Attribute queries

One of the major innovations of GIS technology is the linking of tabular data to the graphic features in map layers. Not only does this linkage lead to easy symbolization of maps through the use of attribute values found in tables, but it also allows unique and powerful attribute queries that zero in on selected locations or events of interest.

Attribute queries are based on the criteria portion of a Structured Query Language (SQL) command. SQL is the de facto standard query language of database packages and many application software packages, including ArcGIS 10. This chapter provides an introduction to SQL; you can find free, in-depth, interactive SQL tutorials on the Internet. A very good one is available at w3schools.com.

A simple SQL criterion has the following form:

```
"attribute name" <logical operator> attribute value
```

"Attribute name" is any attribute column heading or field name in an attribute table. There are several logical operators available, including the familiar ones such as =, >, >=, <, and <=. "Attribute value" is related to the values you seek. For example, the following simple criterion selects all offense crimes that have Uniform Crime Report (UCR) hierarchy values that are less than or equal to 4—that is, all serious violent crimes listed in the hierarchy (see the Hierarchy spreadsheet in the CodeTables folder in Data):

```
"Hierarchy" <= 4
```

The ArcMap implementation of SQL requires that an attribute name, such as Hierarchy, be placed in double quotation marks in the query syntax. After executing a query with the above criterion, ArcMap highlights all the records in the table and all the features on the map (crime points, in this case) that satisfy this criterion.

Compound criteria are made up of two or more simple criteria connected with either an AND or an OR connector. "AND" means that both connected simple criteria must be true for corresponding records and features to be selected. For example, for SQL to select a record for the following multiple-criteria request,

```
"DateOccur" >= date '2008-08-01' AND
"DateOccur" <= date '2008-08-21'
```

the DateOccur value must be greater than or equal to August 1, 2008, and less than or equal to August 21, 2008. So, SQL will select a record with the date August 5, 2008, for example, but it will exclude a record with the date August 31, 2008. The attribute values in the preceding example criteria are in the format ArcMap requires for use in queries, regardless of the specific date format used in the attribute table. The ArcMap Query Builder interface helps you get the values you'll need in queries, as you will see in the exercises in this chapter.

The OR connector can be tricky to use. In common parlance, you might say, "I want all aggravated assaults *and* robberies," but in SQL, the expression is "aggravated assaults *or* robberies." Hence, the use of the OR connector. The reason is that for each offense record, there can only be a single crime type, so it is impossible for an offense record to be both an aggravated assault and a robbery. (Note that a crime incident with the same criminal can have one or more offense records, each with a different crime type. For example, an incident may involve an aggravated assault and a robbery, but it would be composed of two separate records in the database.) So, the proper SQL criteria in this case are as follows:

```
("Crime" = 'Aggravated Assault' OR
 "Crime" = 'Robbery')
```

ArcMap requires that text values be put in single quotation marks, as shown here. Although there are exceptions, you will almost always need to place OR criteria in parentheses as shown if the criteria are combined with other criteria, such as a date range. The reason is, just as with algebraic expressions, logical expressions are executed one at a time, in specific order. For example, SQL executes AND comparisons before OR comparisons, which can result in incorrect information unless you use parentheses to control the order of execution. SQL executes comparisons in parentheses first. A little experience in checking the input data for queries and the output information will guide you in making compound queries.

Spatial queries

Besides using attribute queries to select spatial features, you can also use spatial queries. Using spatial queries involves using existing spatial features to select other features. This capacity of GIS is very powerful and often used in combination with attribute queries to narrow the field of interest to selected features. The spatial features used for selection can be from existing map layers or in new layers created solely for the situation at hand. For example, the buffers used in proximity analysis are created to find features that are within a certain range of existing features. A question that often arises is, What features are near other existing features? Buffers are used to provide the polygons that are used for selection to find those features.

Internet keyword search   **GIS buffer**

A buffer is a polygon created in reference to existing map features. As shown in the figure, buffers can be created for points, lines, or polygons. First, you must specify the desired radius to create the buffer. The buffer of a single point is a circle; the buffer of a line has rounded corners where the buffer radius swings out around end points; and the buffer of a polygon is another polygon, though somewhat smoothed and larger. (Note that although a circular buffer appears to be a circle, it is actually a series of very short straight lines that are connected to form a polygon.)

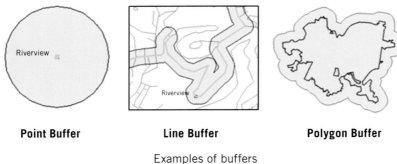

Point Buffer   **Line Buffer**   **Polygon Buffer**

Examples of buffers

You can select features such as crime points within a buffer, and then display or analyze the results. For example, certain crime types such as illicit drug dealing may have the penalties doubled if they occur within a certain distance of schools—say, within 1,000 ft. So, you could screen drug offenses for those that are too close to schools by placing a 1,000 ft buffer around schools, and then using the results to select drug offenses for follow-up review—as well as for more careful measurement. Note that although you can carry out such queries with schools and drug offenses, the results would not have enough accuracy to hold up in court. TIGER (Topologically Integrated Geographic Encoding and Referencing) street centerline maps, available from the U.S. Census Web site and used in this book, do not have sufficient positional accuracy, and the address-matched offense locations have frequent positional errors. Instead, you would need an accurate polygon map with the boundaries of land parcels, including schools, plus a street map with accurate feature locations from a vendor or local government agency, to meet court standards. Regardless, any drug crimes that occurred within specified buffers would need to be verified, perhaps by direct measurement on the ground or by using global positioning system (GPS) technology.

Building and using crime queries

ArcMap makes it simple to build complex queries. ArcMap has an easy-to-use Query Builder interface that allows you to build attribute queries by picking elements from lists and buttons, and then it verifies the correctness of your query syntax. ArcMap makes building spatial queries just as easy. The software provides tools that have computer forms you fill in to help you build a query.

Tutorial 5-1

Creating attribute queries

There are three primary kinds of attribute queries that police need to make to analyze crimes, and these are often combined and used in clever ways. The first and most fundamental kind is attribute queries by crime type and date interval (the "what and when"). Such queries often combine several crime types with the use of logical operators. The second type of attribute query adds criteria such as time of day or day of the week—for example, weekday versus weekend crime (an extension of the "when"). Clearly, crime patterns can differ as a result of the time of day or the day of the week. The third type of attribute query adds criteria based on the attributes of the people (the "who") or the objects (the "what"), such as vehicles, involved in a crime and includes the modus operandi variables (the "how") if available.

You can use queries to select points in a master map layer or to set the definition query properties of a master layer that is used to create layers in a map document. Generally, a point that is selected appears highlighted in the selection color, both in the attribute table and on the map. All crime points in the master map layer remain visible. A definition query, on the other hand, displays only the points that satisfy the query criteria without the need for highlighting. Note that while you briefly use definition queries in chapter 4, this section starts from the beginning and takes you all the way through the process.

Open map document

1 **Open Tutorial5-1.mxd in ArcMap.** The resulting map of Pittsburgh, zoomed to the Police Sector 1-5 bookmark, has only the map layers that are essential for querying—namely, the master map layer (Offenses2008), Streets, and Police Sectors, plus Rivers, because rivers are key geographic features and natural barriers to social and criminal interactions.

2 **Save your map document to your chapter 5 folder in MyExercises.**

Start with date range in selection by attributes

Crime analysis queries almost always use date range criteria, which are often expressed in days. You need to specify the start date and the end date of the query with the use of SQL syntax (or language rules). Given that, the Query Builder interface that ArcMap provides does most of the detailed work for you.

A selection by attributes identifies the subset of features in a map layer that meet the query criteria and so will get the selection color both on the map and in the corresponding attribute table. Once the features are selected, you can do additional processing. For example, you can export the selected features to a new map layer, and then apply tools to that layer to process only the selected features.

1 **From the Menu bar, click Selection > Select By Attributes.** ArcMap uses Offenses2008 as the default map layer for the selection because it is the top layer in the table of contents.

2 **In the large top panel of Query Builder, scroll down, double-click "DateOccur", click the >= button, and click Get Unique Values. Scroll down in the resulting values panel and double-click date '2008-08-01 00:00:00'. Click the And button. Then double-click "DateOccur" and click the <= button. Scroll down in the values panel and double-click date '2008-08-21 00:00:00'. Click Verify and fix any errors.** Compare your query criteria with those in the figure. You can type directly in the criteria panel after the WHERE statement to fix small errors. If necessary, you can place the pointer where needed and type entries, delete parts of entries, or make new selections to rebuild a criterion.

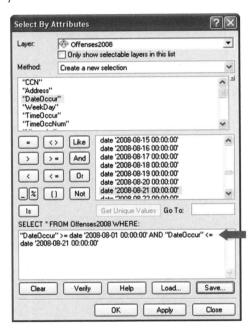

3 **Click OK twice.** All Offenses2008 points remain on the map, but the points that meet the date range criteria are now displayed in the selection color.

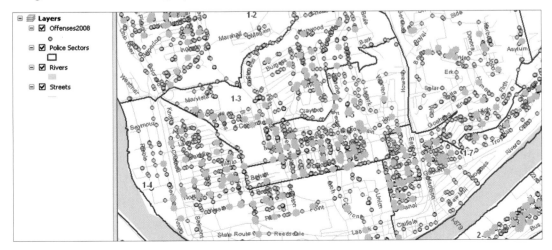

4 **Right-click Offenses2008 and select Open Attribute Table.** At the bottom of the table, a register shows that 2,666 out of 42,108 features are selected.

5 **Click the "Show selected records" button at the bottom of the table.** Now, the table displays only the 2,666 selected records.

	CCN	Address	DateOccur	WeekDay	TimeOccur	TimeOccNum
▶	2008025028	SAW MILL RUN BL E & WHITED ST	8/1/2008	6	12:17:00 AM	1217
	2008025036	352 W CARSON ST	8/1/2008	6	1:20:00 AM	120
	2008025038	6412 MEADOW ST	8/1/2008	6	1:36:00 AM	136
	2008025040	1113 LARIMER	8/1/2008	6	2:00:00 AM	200
	2008025042	6515 FRANKSTOWN AV	8/1/2008	6	2:00:00 AM	200
	2008025051		8/1/2008	6	3:00:00 AM	300

6 **Click the "Show all records" button and close the table.** The selection is temporary in that you can clear it and remove it.

7 **On the Menu bar, click Selection > Clear Selected Features.**

Create date range criteria in a definition query

Although selecting by attributes identifies a subset of features for a map layer, leaving all features visible, a definition query limits the display of a map layer to only those features that satisfy the query criteria. Although definition queries can be easily changed or deleted, the resulting maps, nonetheless, have a more permanent look to them. The same query builder you use to build definition queries is used to select by attributes. Thus, while the remaining attribute queries in this chapter are definition queries, you can use the same logic for both.

1 **Right-click Offenses2008 and select Properties. Click the Definition Query tab, and then click the Query Builder button.**

2 **In the large top panel of Query Builder, scroll down and double-click "DateOccur". Click >=, and then click Get Unique Values. Scroll down in the values panel, double-click date '2008-08-01 00:00:00', and click And. Double-click "DateOccur", click <=, scroll down in the values panel, and double-click date '2008-08-21 00:00:00'. Then click Verify and fix any errors.**

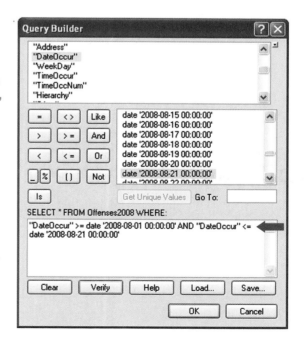

3 **Click OK twice.**

4 **Right-click Offenses2008, click Open Attribute Table, and select any record to get an accurate count of records at the bottom of the table.** You should find that there are 2,666 records in the table. If your query has an error, fix it.

5 **Right-click the DateOccur heading in the table, select Sort Ascending, and scroll down to see that you have records for the first 21 days of August 2008. When you are finished, close the table.** With a good working query in hand, you can save it for future use by just changing the dates.

6   Right-click Offenses2008 in the table of contents and select Properties. Click the Definition Query tab, and then click Query Builder.

7   Click Save, go to your chapter 5 folder in MyExercises, and change File Name to **OffensesDateRange.exp**. Click Save.

8   In the Query Builder window, select the query criterion text in the bottom panel and press DELETE. Now your query is gone, but you can reload it from the stored file.

9   Click Load and double-click OffenseDateRange.exp. Your query is back.

10 Click OK twice.

Add crime types to a query by using a range

Crime types are codes, often entered by using lists in the code tables. The result is that codes are consistently entered with single "official" spellings. Thus, you know that when you query for a crime type, you will get all such crimes that are entered in the system. If you want crimes that are in a sequence using the UCR hierarchy, for instance, it is easy to add a range of numbers by using AND as a connector.

1   Open the Query Builder for Offenses2008.

2   In the bottom panel of Query Builder, put your cursor at the end of the current date-range condition and click And. Scroll down in the top panel, double-click "Hierarchy", and click >=. Then type a blank space and **1**. Click And. Next, double-click "Hierarchy" and click <=. Then, type a blank space and **4**.

Your query should read:

```
"DateOccur" >= date '2008-08-01 00:00:00' AND
"DateOccur" <= date '2008-08-21 00:00:00' AND
"Hierarchy" >= 1 AND
"Hierarchy" <= 4
```

3   Click OK twice.

4   Examine the attribute table for Offenses2008. Note that there are now 213 records, all serious violent crimes with hierarchy numbers 1–4 for the first 21 days of August 2008.

5   Close the attribute table.

6   Save your query as **OffensesDateRangeHierarchyRange.exp** to the chapter 5 folder in My Exercises.

Add crime types to a query by using OR connectors and parentheses

There are often times when you'll want a selection of crime types that do not follow a sequence. In such cases, you can select crimes by using the OR connector and parentheses. OR conditions are tricky, so always check the output of queries you build for the first time to make sure you are getting the desired information.

1 Right-click Offenses2008 and select Properties. Click the Definition Query tab, and then click Query Builder.

2 In the bottom panel of Query Builder, select the SQL expression **"Hierarchy" >= 1 AND "Hierarchy" <= 4** and delete that text by pressing DELETE. The next SQL expression to create follows. You can create it on your own or follow the instructions in step 3. Note that if you do not include the parentheses, the query will yield disorderly conduct offenses within the first 21 days of August 2008 and then *all* vagrancies and vandalisms included in the original table. The parentheses force the three OR conditions to execute first, and then limit selections to the date range.

```
"DateOccur" >= date '2008-08-01 00:00:00' AND
"DateOccur" <= date '2008-08-21 00:00:00' AND
("Crime" = 'Disorderly Conduct' OR
"Crime" = 'Vagrancy' OR
"Crime" = 'Vandalism')
```

3 Place the pointer at the end of the remaining expression and type a blank space and left parenthesis. Scroll down in the top panel, double-click "Crime", and click =. Click Get Unique Values, double-click 'Disorderly Conduct' in the values panel, and click Or. Double-click "Crime" in the top panel, click =, double-click 'Vagrancy' in the values panel, and click Or. Then double-click "Crime" in the top panel, click =, double-click 'Vandalism' in the values panel, and type a right parenthesis.

4 Click Verify and fix any errors. Then click OK twice.

5 Examine the attribute table of Offenses2008. Now, there are 485 records, all disorderly conducts and vandalisms occurring in the first 21 days of August 2008. There were no vagrancies.

6 Save your query as **OffensesDateRangeCrimeSelection.exp** to your chapter 5 folder in MyExercises.

Add day-of-week range to a query

Next, you'll use an attribute, WeekDay, which is a code for the day of the week, where Sunday is 1, Monday is 2, and so on. Generally, crimes vary by day of the week in one of three patterns: on a weekday (Monday through Thursday), on a Friday, or on the weekend (Saturday or Sunday). In the next set of steps, you'll build a query for weekend crimes.

1 Right-click Offenses2008 and select Properties. Click the Definition Query tab, and then click Query Builder.

2 In the bottom panel, delete all text after the date range, but leave the AND connector intact. The query expression you are about to build is as follows:

```
"DateOccur" >= date '2008-08-01 00:00:00' AND
"DateOccur" <= date '2008-08-21 00:00:00' AND
("WeekDay" = 1 OR
"WeekDay" = 7)
```

3 Type a blank space and a left parenthesis. Double-click "WeekDay" in the top panel and click =. Then type a blank space and **1**. Click Or. Next, double-click "WeekDay" in the top panel and click =. Then type a blank space and **7**. Finish by typing a right parenthesis.

4   Click Verify and fix any errors. Then click OK twice.

5   Examine the attribute table of Offenses2008. Now, there should be 743 records, all weekend days within the first 21 days of August 2008.

6   Save your query as **OffensesDateRangeWeekEnd.exp** to your chapter 5 folder in MyExercises.

Add time-of-day range, including midnight, to a query

It is easy to add a time-of-day criterion as an interval to a query by using the numerical time equivalent, TimeOccNum, which appears as an attribute in Offenses2008. This attribute is similar to military time but has the integer data type—for example, 2:35 A.M. gets the value 235, and 2:35 P.M. gets the value 1435. The only difficult case is when the desired time interval includes values on either side of midnight, and that is the one you build here. Suppose the desired time interval is 10:00 P.M. to 2:00 A.M. The solution is to use two intervals connected by OR—from 10:00 P.M. to 11:59 P.M. (2200 to 2359) or from 12:00 A.M. to 2:00 A.M. (0 to 200).

1   Right-click Offenses2008 and select Properties. Click the Definition Query tab, and then click Query Builder.

2   In the bottom panel, delete all text after the date range, but leave the AND connector intact. The query expression you are about to build is as follows:

```
"DateOccur" >= date '2008-08-01 00:00:00' AND
"DateOccur" <= date '2008-08-21 00:00:00' AND
(("TimeOccNum" >= 2200 AND
"TimeOccNum" <= 2359) OR
("TimeOccNum" >= 0 AND
"TimeOccNum" <= 200))
```

Notice the outside parentheses combining the two OR conditions.

3   Type a blank space and two left parentheses. Double-click "TimeOccNum" in the top panel and click >=. Then type a blank space and **2200**. Click And. Next, double-click "TimeOccNum" in the top panel and click <=. Then type a blank space and **2359**. Type a right parenthesis and click Or. Next, type a blank space and a left parenthesis. Double-click "TimeOccNum" and click >=. Then type a blank space and **0**. Click And. Next, double-click "TimeOccNum" and click <=. Then type a blank space and **200</**. Finally, type two right parentheses.

4   Click Verify and fix any errors. Then click OK twice.

5   Examine the attribute table of Offenses2008. Now, there are 677 records. The results include all crimes that occurred within the date interval and within the continuous time interval that runs from 10:00 P.M. to 2:00 A.M.

6   Save your query as **OffensesDateRangeTimeOfDayAcrossMidnight.exp** to your chapter 5 folder in MyExercises.

Add person attributes to a query

Designing queries for persons or objects requires a lot of ingenuity and detective work. For example, suppose an officer suspects that an unsolved crime was committed by a male youth, of non-Caucasian race, who resides on a street that contains the word "wood" it its name, such as Woodland or Pinewood. Often, you will need to search parts of text values, such as part of an address, to get the one you want, and the LIKE logical operator provides the solution. The LIKE operator allows you to use a wildcard character, %, in text values to represent zero, one, or more characters. For example, '%WOOD%' finds records with the corresponding values of Woodland, Pinewood, and Oakwoodland.

Here, you use several criteria, including age, race, and a partial clue about street of residence, to look for suspects.

1   Right-click Offenses2008 and select Properties. Click the Definition Query tab, and then click Query Builder.

2   In the bottom panel, delete all text after the date range, but leave the AND connector intact. The query expression you are about to build is as follows:

```
"DateOccur" >= date '2008-08-01 00:00:00' AND
"DateOccur" <= date '2008-08-21 00:00:00' AND
"ArrSex" = 'M' AND
"ArrAge" <= 18 AND
"ArrRace" <> 'W' AND
"ArrResid" LIKE '%WOOD%'
```

Of interest here are the "not equal to" (<>) logical operator, the LIKE logical operator for text values, and the wildcard character (%) that represents and replaces zero, one, or more characters. Text query values are case sensitive, so "wood" needs to be in all caps (WOOD) because attribute values for ArrResid are in all caps. The wildcard expression must also appear in single quotation marks because it is a text value.

3   Double-click "ArrSex" in the top panel and click = Click Get Unique Values, double-click 'M' in the values panel, and click And. Next, double-click "ArrAge" and click <=. Then type a blank space and **18**. Click And. Next, double-click "ArrRace" and click <>. Click Get Unique Values, double-click 'W,' and click And. Then double-click "ArrResid" and click LIKE. Finally, type a blank space and **'%WOOD%'**.

4   Click Verify and fix any errors. Then click OK twice.

5   Investigate the attribute table. This attribute query for persons narrowed the search to just a few suspects for officers to investigate. Because this query is for a unique situation, you won't have to bother to save it for reuse.

ArrFName	ArrLName	ArrRace	ArrSex	ArrResid	ArrLoc	ArrDate	ArrAge
▸ Daryl	Joseph	U	M	817 INWOOD ST		8/10/2008	18
Fred	Soto	B	M	1068 WOODLOW ST	1048 WOODLOW ST	8/5/2008	18
Guy	Cantu	B	M	2412 WOODSTOCK AV	616 COLLIER ST	8/13/2008	17

6   Close the attribute table and save your map document.

Tutorial 5-2

Creating spatial queries

One of the biggest uses of spatial analysis is to find crimes that are committed near selected points or lines on a map. For example, the Pittsburgh nuisance bar law allows public officials to close bars that generate or attract significant amounts of crime by finding concentrations of crimes that occurred nearby. In this case, selected crimes within a block or two of a target bar provide evidence of a bar's contribution to crime. Another example is attempting to identify possible car thefts by a suspected car thief who steals cars for transportation. Routine-activity theory suggests looking along streets connecting the thief's residence, workplace, and girlfriend's (or boyfriend's) residence—all of the thief's anchor points. Using a map of buffered, anchor-point routes that include car theft and recovery locations can lead to higher clearance rates of such crimes.

Open map document

Some public officials and residents in Pittsburgh believe that the South Side neighborhood has too many bars and that its concentration of bars attracts and generates crimes. For example, bars are notorious locations for drug dealing, and some drug users support their habit by robbing people. So in this tutorial, you analyze drug and robbery offenses in the vicinity of South Side bars. A good way to study crimes in the vicinity of bars is to build circular buffers around the bars, and then use those buffers to select crimes. The crimes selected, then, occur in proximity to the bars.

1   **Open Tutorial5-2.mxd in ArcMap.** The starting map has police sectors, rivers, and streets for location; the offenses2008 master layer with a query definition for summer 2008 and drug or robbery offenses; and a point map layer for the 27 bars on the South Side. The definition query for offenses is `"DateOccur" >= date '2008-06-01 00:00:00' AND "DateOccur" <= date '2008-08-31 00:00:00' AND ("Crime" = 'Robbery' OR "Crime" = 'Drug Violations')`. Clearly, these crimes occur near bars on the South Side. You can use buffers to analyze the crimes.

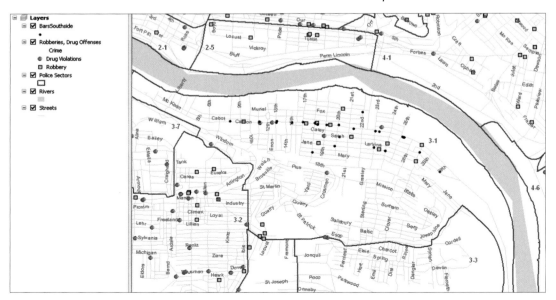

2   **Save your map document to your chapter 5 folder in MyExercises.**

Create bar buffer

Block lengths average 275 ft on Pittsburgh's South Side, so you can use this radius to generate a buffer around each bar.

1 From the Menu bar, click Geoprocessing > Buffer.

2 Enter selections in the Buffer window as shown in the figure.

3 Click OK > Close.

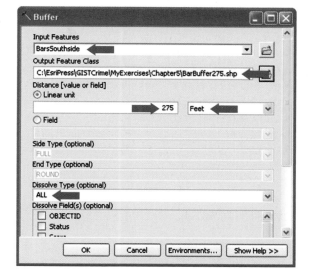

4 Right-click BarBuffer275 and select Properties. Then click the Display tab.

5 Type **50** in the Transparent box and click OK. Now, you can see the streets below the 50% transparent buffers. Clearly, many drug violations and robberies occur near bars on the South Side.

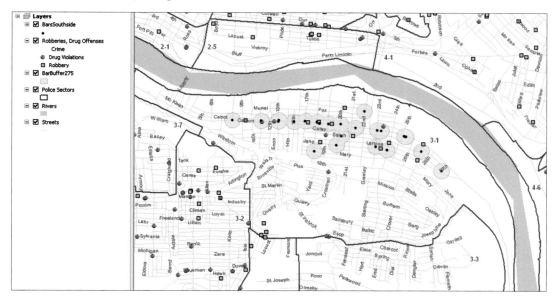

Obtain statistics on crimes within a buffer area

A good question to ask is, What proportion of drug violations and robberies in Police Sector 3-1 (the South Side's police sector) are within roughly a block of a bar? This is easy to determine by using selection. First, you find the number of such crimes in the police sector, then determine how many of these crimes occur within the buffer, and finally calculate the percentage of crimes that are being committed within a block of a bar.

1   In the table of contents, right-click Police Sectors, select Open Attribute Table, and click the row selector for sector 3-1 to select it. Then close the table.

2   From the Menu bar, click Selection > Select By Location and enter selections as shown in the figure.

3   Click OK.

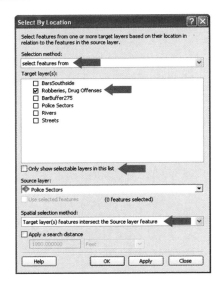

4   Open the Robberies, Drug Offenses attribute table and notice that 26 records are selected. Then close the table.

5   Click Selection > Clear Selected Features.

YOUR TURN   YOUR TURN   YOUR TURN   YOUR TURN   YOU

Use Selection to select drug violations and robberies within the bar buffer. You should find 11 such crimes. Thus, 11 out of 26 drug violations and robberies, or 42% of all such crimes in Police Sector 3-1, are within roughly a block of a bar. That does not mean, of course, that the bars are necessarily causing these crimes, but clearly the bar area merits police attention. When you are finished, close the table and clear all selected features.

Create a multiple-ring buffer

Sometimes, it is helpful to add rings to a buffer to investigate the drop-off in crime in relation to the distance from crime-prone land uses. So next, you'll use a multiple-ring buffer to add those rings.

1   From the Menu bar, click Geoprocessing > ArcToolbox.

2   In ArcToolbox, expand the Analysis toolbox and then the Proximity toolset. Then double-click the Multiple Ring Buffer tool.

3   Enter selections as shown in the figure (for Distances, type one value at a time and press the Add button to add to the list of distances).

4   Click OK > Close. Then close ArcToolbox.

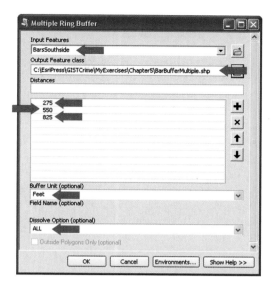

5   In the table of contents, right-click BarBufferMultiple and select Properties. Click the Display tab and type **50** in the Transparent box. Click OK.

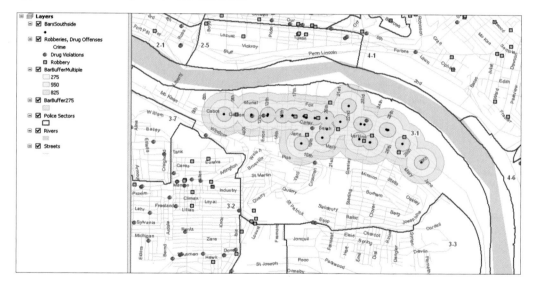

6   Right-click BarBufferMultiple again and select Open Attribute Table. Notice that the resulting buffer has three polygons. One is the original circular buffer, and the other two are rings around the buffer.

YOUR TURN   **YOUR TURN**   YOUR TURN   YOUR TURN   YOUR

You already know that there are 11 drug violations and robberies within the 275 ft buffer. Select the BarBufferMultiple polygon for 550 ft, and then use Selection to find the number of such crimes that occurred within that ring (roughly one to two blocks from a bar). You should find that there are five crimes. Do the same for the 825 ft ring. You should find one crime. So, most of the drug and robbery crimes associated with bars take place within two blocks of a bar. Also, the one- and two-block buffers account for 17 of the 26 crimes (65%). When you are finished, clear selected features, and then save and close your map document.

Assignment 5-1

Analyze leading-indicator crimes by day vs. by night

Leading-indicator crimes can help forecast large increases in the crimes they predict (Cohen et al. 2007; Gorr 2009). For example, disorderly conduct, vagrancy, and vandalism crimes tend to lead serious property crimes. So, if there is a spike in leading-indicator crimes in an area in the previous month, there is a good chance serious property crimes, such as larcenies and burglaries, will spike the following month in the same area. Broken-windows theory suggests that certain behaviors such as vandalism (i.e., broken windows) can signal impending, more serious crime—namely, that soft crimes tend to harden over time into more serious crimes as perpetrators establish themselves in an area. If such leading behavior is present and detectable, police can be proactive and perhaps prevent serious violent crimes in the community.

For this assignment, investigate the patterns of leading-indicator crimes by day versus by night.

 Internet keyword search   **broken-windows theory**

Create new map document

Create a new map document called **Assignment5-1YourName.mxd** and save it to your assignment 5-1 folder in MyAssignments. Use relative paths for your map document. Add the following map layers and data:

+ Standard spatial context layers: Streets, Rivers, ZoningCommercialBuffer, and Neighborhoods from the Pittsburgh geodatabase in the Data folder.

+ Two copies of Offenses2008 from the Police geodatabase in the Data folder. For the purposes of this assignment, leading-indicator crimes are disorderly conduct, drug violations, liquor law violations, and simple assault.

Symbolize layers

+ Label one copy of Offenses2008 as **Leading Indicators Day** in the table of contents and the other as **Leading Indicators Night**. Symbolize each layer with a single symbol to indicate day versus night. Note that you will not distinguish crime offenses by crime type, but simply by day versus night.

+ Symbolize Streets, Rivers, ZoningCommercialBuffer, and Neighborhoods using best practices, including labels where relevant. You will use this map while zoomed in to see several neighborhoods, or even closer in. Use no color fill for noncommercial areas and a color fill of 50% transparent for commercial areas (Commercial = 1).

Query leading-indicator crimes

You need to create two similar definition queries, one for daytime leading indicators and the other for nighttime leading indicators. Make each query for the month of July 2008. To query by time of day, use the TimeOccNum attribute in Offenses2008. Its virtue is that it sorts correctly in ascending order by time of day: for example, 100 = 1:00 A.M., 630 = 6:30 A.M., 1159 = 11:59 A.M., 1700 = 5:00 P.M., and 2459 = 12:59 A.M. Define days as 7:00 A.M. to 7:00 P.M. (including, as is customary, the left side of the interval up to but not including the right side) and the balance of the day as night.

Hint: Enclose each group of criteria of the same type in parentheses. For example, include the two simple criteria that define the date range for July 2008 in parentheses. Also, when you have completed a query, check the resulting attribute table to make sure the desired records are output from each of the criteria; for example, sort tables by TimeOccNum and make sure the correct times are included.

Analyze leading indicators

Obtain frequency counts of leading-indicator crimes for all of Pittsburgh by crime type and by day versus by night. Create a table that has frequency counts and percentages of crimes by crime type for days and for nights. Create a table that has the frequencies in a Word document called **Assignment5-1YourName.docx** and save it to your assignment 5-1 folder. Include the title, your name, and the date at the top of your document and write a few paragraphs on your findings.

Hint: ArcMap does not have the functionality to tabulate frequencies as statistical output, but it does tabulate and display frequencies if you symbolize by Categories > Unique Values. So, use the following "trick": Temporarily resymbolize Leading Indicators Day using Crime as the value field in Categories. Add all values to the major panel. Under the Count column of the results, you will find the desired frequencies, which you can copy and paste into a Word document. Cancel the Layer Properties window so as not to modify your map. Use the same process for Leading Indicators Night.

Zoom in on the portion of Pittsburgh that includes the following neighborhoods (see the attribute table for Neighborhoods to locate the neighborhoods):

+ CBD—the central business district of downtown Pittsburgh
+ Strip District—so named because it is a narrow strip of land that originally had fresh-produce warehouses and distribution centers but now has many fresh-produce markets, bars, and restaurants
+ Bluff—the area that has Duquesne University
+ Crawford-Roberts—an area that has new, mixed-income housing with both single- and multiple-family dwellings
+ Bedford Dwellings, Middle Hill, and Terrace Village—all with elements of poverty

Export the resulting map as an image, or if you know how to create a map layout from chapter 3, do so and export it as an image called **Assignment5-1YourName.jpg** to your assignment 5-1 folder. Include the map in your Word document and describe the major patterns you see.

What to turn in

Use a compression program to compress and save your assignment 5-1 folder to **Assignment5-1YourName.zip**. Turn in your compressed file.

Assignment 5-2

Analyze robberies near check-cashing businesses

From routine-activity theory, it follows that criminals, including robbers, need good targets in areas or settings that have low guardianship and, hence, offer a lower chance of getting caught. Check-cashing establishments and areas around them generally meet these conditions. These businesses serve mostly poor people who have low credit ratings and no bank accounts, and thus, such businesses generally are located in run-down commercial areas. Guardianship from shopkeepers and citizens is generally low in poor areas, and check-cashing customers leave the premises with cash, making them good targets for thieves.

For this assignment, use proximity analysis of check-cashing stores in Pittsburgh to investigate robberies. Of course, not every robbery in the vicinity of a check-cashing establishment has a check-cashing client as its victim.

Create new map document

Create a new map document called **Assignment5-2YourName.mxd** and save it to your assignment 5-2 folder in MyAssignments. Use relative paths for your map document and use state plane for Southern Pennsylvania (NAD83 US Feet) as the data frame coordinate system. Add the following map layers and data:

+ Standard spatial context layers: Streets, Rivers, and Tracts from the Pittsburgh geodatabase.

+ Two point feature layers: Offenses2008 from the Police geodatabase and CheckCashingPlaces from the Pittsburgh geodatabase. Add two copies of Offenses2008. Note that for this data, WeekDay = 1 is Sunday, 2 is Monday, and so on.

+ Data on poverty by tract: AllCoPoverty$ sheet from the Poverty spreadsheet in Downloads of the chapter 8 folder in FinishedExercises.

Symbolize layers

+ Keep the two copies of Offenses2008 turned off for now.

+ Symbolize Streets, Rivers, and CheckCashingPlaces using best practices, including labels where relevant.

+ Join Tracts using its CTIDFP00 tract identifier to the Tract attribute in AllCoPoverty$ from the Poverty spreadsheet. Symbolize tracts, using a gray color ramp for PopPov normalized by Pop (which yields the fraction of the population living in poverty) and five quantiles. Make the fill color black for the highest quantile.

Query robberies

Check-cashing businesses are open 9:00 A.M. to 7:00 P.M., Monday through Saturday. Assume that robberies in the vicinity of check-cashing places are more likely to occur in warm-weather months (May through September). Turn one of your Offenses2008 layers on and create a definition query to yield robberies for May through September. Symbolize with a size 6, red Circle 1 point marker. Rename this layer **Warm-Weather Robberies**.

Move your other Offenses2008 layer in the table of contents above the layer you just symbolized. Include criteria just used, plus use WeekDay to restrict robberies to the six days per week when check-cashing establishments are open and use TimeOccNum to restrict robberies to the hours when check-cashing establishments are open. TimeOccNum sorts correctly in ascending order by time of day; for example, 100 = 1:00 A.M., 630 = 6:30 A.M., 1159 = 11:59 A.M., 1700 = 5:00 P.M., and 2459 = 12:59 A.M. Rename this layer **Warm-Weather Robberies Workday** and symbolize this layer with a size 6, orange Circle 1 point marker. The orange point markers are the most relevant subset of the red point marker robberies for check-cashing businesses. The visible red point markers, not covered up by orange point markers, are robberies that are not associated with check-cashing establishments and should form different patterns.

Buffer check-cashing establishments

Create 300, 600, and 900 ft multiple-ring buffers for the check-cashing establishments. Make the resulting buffers 50% transparent.

Analyze robberies

Zoom in on two check-cashing establishments, one at a time, that you feel merit further analysis in regard to associated robberies. Produce a Word document called **Assignment5-2YourName.docx** and save it to your assignment 5-2 folder. Include a map image from each establishment and a table of robberies with relevant attributes. Include the title, your name, and the date. Create layouts with titles and legends. Write a few sentences about each establishment and proximate robberies. Copy and paste the SQL code of your query that produced the robberies of interest.

What to turn in
Use a compression program to compress and save your assignment 5-2 folder to **Assignment5-2YourName.zip**. Turn in your compressed file.

References

Cohen, J., W. L. Gorr, and A. Olligschlaeger. 2007. Leading indicators and spatial interactions: A crime forecasting model for proactive police deployment. *Geographical Analysis*. 39 (1): 105–27.

Gorr, W. L. 2009. Forecast accuracy measures for exception reporting using receiver operating characteristic curves. *International Journal of Forecasting*. 25 (1): 48–61.

Chapter 6

Building crime map animations

Map animation, which involves displaying a set of crimes in a sequence of map frames, is a relatively new crime analysis tool. Map animation has the potential to unlock the dynamics of crime patterns, giving crime analysts a vital tool. For example, police can catch serial criminals "in the act" by studying and learning their regular patterns and habits, and then staking out the likely location for the next crime in a series. A map animation is a powerful tool for discovering such regularities. Likewise, animation of pin maps can show patterns of emerging, persisting, and declining or extinguished hot spots. Using the tools in ArcMap makes map animation a good, practical tool for crime analysis.

Crime map animation concepts

Map animation consists of a series of map frames or snapshots in which spatial context features such as streets are fixed over time, while the locations of crime change sequentially. A crime map animation is an exploratory tool for studying the dynamics of crime patterns that can also be used as a presentation tool for displaying the patterns you identify as an analyst. Map animation is a relatively new tool for spatial data analysis that offers promise in many fields of application.

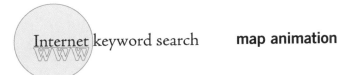 Internet keyword search   **map animation**

Crime map animation principles

Crime map animation generally fixes the map extent and spatial context features, such as streets and police sectors, in place while allowing the crime locations to vary over time. The result, however, is often not very informative—just a lot of crime points jumping around—unless you know the "secrets" of creating useful crime animations: (1) you have to provide a *crime spatial context* similar to the one used for the field officers' pin map; and (2) you have to annotate patterns of interest in the animation to draw the viewer's attention to them.

For example, for hot spot analysis, the idea is to display some historical crime points along with the new crime points so the observer can see if the new crimes are part of a persisting hot spot, are isolated events, or are emerging as a new hot spot. It is also possible to determine whether an old hot spot is disappearing. Once crime patterns are identified, the crime analyst needs to annotate the animation with circled areas and labels to direct the viewer's attention. Good design principles make crime map animations pay off as a useful tool for crime analysis.

Integrated crime mapping and animation

It was once quite time consuming to create a crime map animation. You had to create each frame separately (that is, configure a separate map document for each frame with selected crime data and export an image manually each time), and then assemble the frames in an external software package. The powerful Animation toolbar in ArcMap makes map animation a quick, effective tool for crime analysis. It automates map frame production, making crime animation fast and straightforward.

Building crime map animations

In this chapter, you use the Animation toolbar in ArcMap to build two kinds of animation. In tutorial 6-1, you learn how to build animations for serial crimes. In tutorial 6-2, you learn how to build animations for detecting and studying crime hot spots.

Tutorial 6-1

Building an animation for serial crimes

The goal of an animation for serial crimes is to identify likely locations where a serial criminal might strike next. The criminal, in this case, robbed ATM users of their cash and valuables over a period of about a month, often committing several robberies on the same day, while switching target areas within a few neighborhoods. Pittsburgh Police Bureau crime analysts mapped these offenses and learned that the robber would return to previous crime spots, so they staked out ATMs in the robber's "hunting grounds" and were able to catch him. In this exercise, you build an animation that shows the robber's progression of robberies to serve as an aid in apprehension.

Open map document

1   **Open Tutorial6-1.mxd in ArcMap.** The map document has the familiar police sectors, rivers, and streets, plus the series of 21 robberies (ATMRobberies). The labels on the map are the sequence numbers in the series of robberies. You will be using the ATMCumulative layer later to provide a temporal crime context.

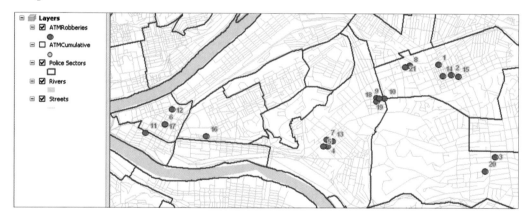

2   Save your map document to your chapter 6 folder in MyExercises.

Set time properties for a map layer

Time properties for a map layer such as a crime layer provide a lot of information that helps in the creation of animation frames. Indeed, once the time properties of a map layer are set, just a few additional steps are needed to complete a simple animation.

1   In the table of contents, double-click
    ATMRobberies, click the Time tab, and enter
    selections as shown in the figure, but do not
    click OK.

2   Click the Calculate button.

3   Click OK. There are no apparent changes, but
    now ArcMap has the information it needs to
    animate the ATM crimes.

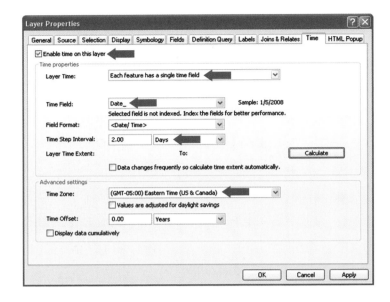

Create time animation

The next step is to create a simple animation.

1   From the ArcMap Menu bar, click Customize > Toolbars > Animation.

2   On the Animation toolbar, click the Animation arrow and select Create Time Animation. Click OK.

3   Click the Animation arrow again, select Animation Manager, and click the Keyframes tab. By default,
    ArcMap provided an animation frame for every two days, but you actually need a new frame for
    every day. So, next, you'll need to update the frame interval from two days to one day.

4   Change the Interval value from 2 to 1 for Time Keyframe 1. Then click the Update button.

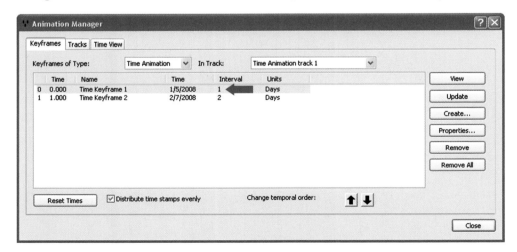

5   Do the same for Time Keyframe 2 and leave Animation Manager open.

View starting crime series animation

There are two ways to view a time animation. First, you can use the Time View, which allows you to view single frames that you choose individually. This option is good for in-depth study. The second, and primary, option is to play the frames as a movie. In this exercise, you learn how to use both options.

1   Click the Time View tab in Animation Manager, and then click anywhere *under* the red horizontal line at its midpoint, 0.500. Clicking there creates a red vertical slider that you can drag to any day on the time animation track. The Time Track view provides the date—in this case, January 21, 2008—and the map shows the ATM robberies committed on that date. Notice that there were five robberies committed in four distinct areas on January 21—one of the robber's most active days.

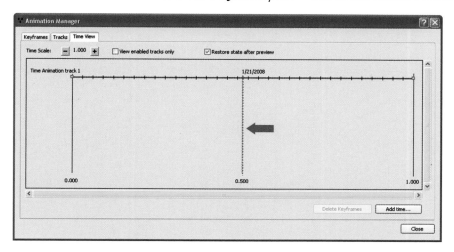

2   Drag the slider to other dates and view the map.

3   Close Animation Manager.

4   On the Animation toolbar, click the Open Animation Controls button ⏯, and then on the Animation Controls toolbar, click the Options button.

5   Click the "By number of frames" option, click Calculate (you will get 34 frames), and change the frame duration to **1.0** second per frame. Now you are ready to play the movie version of the animation.

6   On the Animation Controls toolbar, click the Play button ▶. You can see that the robber was clever, moving around and skipping days.

7   Close the Animation Controls toolbar.

Add temporal context to a crime series animation

To provide a temporal crime context, you can add time properties to a second copy of the ATM crime series, ATMCumulative, and leave the points displayed, cumulatively, as time passes.

1   In the table of contents, right-click ATMRobberies and select Label Features to turn the robbery labels off for this layer. ATMCumulative has the same labels and they are already turned on.

2   Turn ATMCumulative on.

3   Double-click ATMCumulative, click the Time tab, and enter selections as shown in the figure.
    Hint: Be sure to select the "Enable time on this layer" check box.

4   Click Calculate.

5   Click OK.

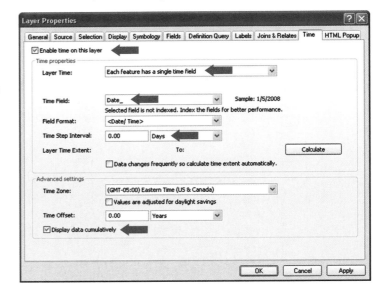

6   On the Animation toolbar, click the Animation arrow and select Clear Animation.

7   Click the Animation arrow again and select Create Time Animation. Click OK.

8   Click the Animation arrow a third time, select Animation Manager, and click the Keyframes tab. ArcMap again provided an animation frame every two days, but you actually need a new animation frame every day. You can now update the frame interval from two days to one day.

9   Change the Interval value from 2 to **1** for Time Keyframe 1 and click the Update button. Then do the same for Time Keyframe 2.

10   Save your map document.

N YOUR TURN   **YOUR TURN**   YOUR TURN   YOUR TURN   YOU

> Explore the new animation using the Time View slider. Also, play the animation. What patterns can you find in the robber's activities? The police were able to catch him at the last crime point in the series. When you are finished, close the Animation Controls toolbar and save your map document.

Use a spreadsheet to create a date stamp (optional exercise)

If you skip this exercise, you can find the end product, ATMDateStamp.csv, in the chapter 6 folder in FinishedExercises. In this exercise, you learn how to create an XY file that will serve as a new point layer with date labels, so the animation can display a time stamp.

1 From the Windows Start menu, click Microsoft Office Excel.

2 Type **X** in cell A1, **Y** in cell B1, and **Date** in cell C1. The values in row 1 of a table become the attribute names of the data you will enter.

3 In ArcMap, move the pointer to a spot in the center of the map above the crime series points. Then type that point's x-coordinate and y-coordinate, rounded up, which are shown on the ArcMap Status bar, into cells A2 and B2 of the Excel spreadsheet; and type the start date of

	A	B	C	D
1	X	Y	Date	
2	1353414	415709	1/5/2008	
3				

the crime series, **1/5/2008**, in cell C2. You can use the coordinates shown in the figure, if you like. The figure also shows the selection box and the "handle" you will use in step 5.

4 Click in cell A2, press SHIFT, and then click in cell C2. This step creates the selection box and the small square handle in the lower-right corner that you will use for dragging.

5 Drag the handle of the selection box to row 35, and then release the handle. Excel is smart! It simply copies the x- and y-coordinates but assumes you want to create a time series of dates. The resulting date range is January 5 through February 7, 2008, matching the crime series date range. If Excel places a series of hash marks (#) in the date cells, it means a wider column width is needed to display the

	A	B	C
1	X	Y	Date
2	1353414	415709	1/5/2008
3	1353414	415709	1/6/2008
4	1353414	415709	1/7/2008
5	1353414	415709	1/8/2008
6	1353414	415709	1/9/2008
7	1353414	415709	1/10/2008
8	1353414	415709	1/11/2008

actual date values. Double-click the right boundary of the column C heading and Excel widens column C just enough to display all the values.

6 Click the Office button and click Save As. Go to your chapter 6 folder in MyExercises, change the type to CSV, and type **ATMDateStamp.csv** for File Name. Click Save > OK > Yes. Close Excel and click No.

Add XY layer

Now, you can add and symbolize the XY layer created in the previous exercise. If you skipped the previous exercise, you can find the needed XY file, ATMDateStamp.csv, in the chapter 6 folder in FinishedExercises. When you add this file, ArcMap displays a warning that there are only a few things you can do with an XY layer (as compared with a full-fledged feature class or shapefile), but that is OK for the purposes of animation.

1 On the ArcMap Standard toolbar, click Add Data. Then click the Browse button, go to your chapter 6 folder in MyExercises, and double-click ATMDateStamp.cvs.

2 From the Menu bar, click File > Add Data > Add XY Data. Then click OK twice.

3   In the table of contents, double-click ATMDateStamp.csv Events, click the Symbology tab, and click the Symbol button. Change the symbol to Circle 1, Size 1, No Color. Click OK. You do not want to see the symbol, but only the label, which you will take care of in the next step.

4   Click the Labels tab of the Layer Properties window and make selections as shown in the figure.

5   Click OK. The map display may show two dates, but it will have only the current day's date when the simulation runs.

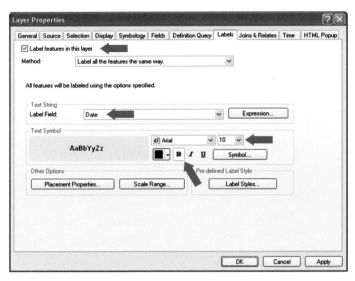

Go through all the steps of creating a new time animation. Add time properties to the ATMDateStamp.csv Events layer. **Hint:** Do not select the "Display data cumulatively" check box. Clear the animation, rebuild it, and change the keyframes to a one-day interval. When ready, play the animation and study it. Finally, save your map document.

Save animation to a file

An animation can be saved to an ArcMap file for editing and later reuse.

1   On the Animation toolbar, click the Animation arrow and select Save Animation File.

2   In the Save Animation window, go to your chapter 6 folder in MyExercises and change File Name to **ATMRobberies.ama.** Click Save. Later, you can load the animation and do more work to it or modify it by clicking the Animation arrow and selecting Load Animation File.

Export animation to a video file

You may have noticed that when you play an animation from the Animation Controls toolbar, the labels have an annoying flicker as each frame changes. If you save the animation as a stand-alone animation file, which you learn how to do in this exercise, the flickering stops. This is reason enough to create an animation file, but in addition, you will have a version of the animation that others can play, using any animation viewer.

1   On the Animation toolbar, click the Animation arrow and select Export Animation.

2 In the Export Animation window, click the Browse button, go to your chapter 6 folder in MyExercises, and change File Name to **ATMRobberies.avi**. Click Options and select the "Enable Off-Screen recording" check box. Click OK > Export > OK. The progress field on the Status bar shows how much of the process is completed. Exporting takes a few minutes.

3 Open a My Computer window, go to your chapter 6 folder in MyExercises, and double-click ATMRobberies.avi. If you have Windows Media Player or some other package for playing videos, your video should play. Notice that the animation plays smoothly and that the point labels do not flicker as they do in ArcMap.

4 Close your media player.

5 Save your map document.

Tutorial 6-2

Building an animation for hot spots

This animation tutorial also applies the principle of temporal crime context introduced in chapter 3. First, you learn how to build a pin map animation without a time interval and see that the result is not informative when it fails to reveal crime hot spots and their dynamic behavior. Then you learn how to add temporal crime context, a date stamp, and a third aid for communicating crime patterns—namely, cluster annotation with an arrow and label directing the user to an emerging hot spot. This tutorial includes several in-depth Your Turns in which you repeat the standard steps for building animation from the first tutorial in this chapter.

Open map document

1 Open Tutorial6-2.mxd in ArcMap. The map is similar to the field officers' pin map in chapters 3 and 4. Serious Crimes Current Hierarchy displays the current day's crimes while Serious Crimes 4 Weeks Context Hierarchy shows the prior four weeks' crimes. The gray and black point markers (which will not be visible on the map until you run the animation) provide sharp contrast with the current day's large and colorful point markers. The map has definition queries for crime type (type 1, serious crimes), but no date limitations because the animation will set them as it creates frames.

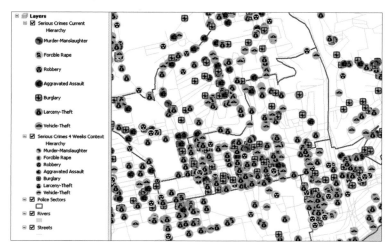

2   Save your map document to your chapter 6 folder in MyExercises.

Create new date column for animating a window of crime points

Displaying a moving window with several days of crime points requires start and end dates. The solution provided in this exercise is to create a new date column for the end date by adding 28 days to DateOccur. Serious Crimes Current Hierarchy and Serious Crimes 4 Weeks Context Hierarchy are layers based on definition queries of Offenses2008.

1   In the table of contents, right-click Serious Crimes 4 Weeks Context Hierarchy and select Open Attribute Table.

2   Click Table Options > Add Field Type. Type **EndDate2** for Name and select Date for Type. Click OK. Note that EndDate already exists with the correct values, but here you are creating EndDate2 with the same values for practice. In the next exercise, you'll use the original EndDate in the animation.

3   Scroll to the right in the table, right-click the column heading for EndDate2, and select Field Calculator.

4   Double-click DateOccur in the Fields panel and click the + button. Type a blank space and **28** to yield [DateOccur] + 28 for the EndDate expression.

5   Click OK. The resulting values for EndDate2 are dates that are 28 days later than DateOccur.

6   Close the attribute table.

N YOUR TURN   **YOUR TURN**   YOUR TURN   YOUR TURN   YOU

Build the starting pin map animation, following the steps of tutorial 6-1:

- Turn off the Serious Crimes 4 Weeks Context Hierarchy layer.

- Add time properties for Serious Crimes Current Hierarchy, including DateOccur in the Time field, Days for Time Step Interval, and whatever step size ArcMap chooses. Click to clear the "Display data cumulatively" check box. Then click Calculate.

- Use Animation Manager to create a time animation and modify Keyframes to an Interval of **1** day.

- On the Animation Controls toolbar, click Options, select the "By number of frames" option, and click Calculate. **Hint:** You should get 366 frames. Type **0.5** for Frame Duration, select the "Play only from" check box, and type **242** in the From frame and **274** in the To frame.

- Play the animation.

You should be unimpressed because about all you can learn from this animation is that crime points jump around. To make this animation informative, you'll need to add a crime temporal context and annotation for emerging hot spots.

Add crime temporal context track to the animation

The next objective is to display a moving window for the Serious Crimes 4 Weeks Context Hierarchy layer with a length of 28 days for each frame. Displaying the moving window of crime points requires a start date and an end date. You created the new date column, EndDate2, in the previous exercise, and can now put its equivalent, EndDate, to work in this exercise.

1 In the table of contents, turn Serious Crimes 4 Weeks Context Hierarchy on.

2 Double-click Serious Crimes 4 Weeks Context Hierarchy, click the Time tab, and select the "Enable time on this layer" check box. Make selections as shown in the figure (note the new Layer Time selection). Be sure to click Calculate to get the value for Layer Time Extent.

3 Click OK.

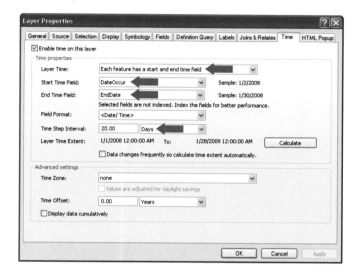

YOUR TURN **YOUR TURN** YOUR TURN YOUR TURN YOUR

Rebuild the animation by using the following steps:

- Clear the previous animation.

- Create a time animation.

- Use Animation Manager to modify Keyframes to an Interval of **1** day.

- On the Animation Controls toolbar, click Options and change Frame Duration from 0.5 to **1**.

- Play the animation.

Use the Time View window of Animation Manager to convince yourself that a given frame of the animation has the date's current crimes, plus a 28-day window of previous crimes. Move the red vertical slider to about the middle of the year and use the Identify tool to see the dates of both current points and context points. **Hint:** You should see dates that are the current date for Serious Crimes Current Hierarchy and dates between the current date and 28 days prior for the context layer.

The context layer helps to identify crime clusters. Next, you'll add a date stamp. Instead of having to create the layer, as you did for the crime series, it is already available as an XY file called PinMapDateStamp.csv in the chapter 6 folder in FinishedExercises.

(continued on next page)

- Add PinMapDateStamp.csv as XY data to the map document.
- Symbolize PinMapDateStamp.csv Events with no color and add a label for Date in size 10, Bold.
- Add appropriate time properties.
- Rebuild the animation again.
- Play the animation.

Explore emerging hot spot

Now that you have a date stamp along with the two crime layers, you are in good shape to use the Time View window of Animation Manager to study spatial crime patterns and look for hot spots. The goal in practice is to identify an emerging hot spot as soon as possible so it can be nipped in the bud.

1 **Click the Time View tab of Animation Manager and move through the July and August time frames.** A hot spot emerges in the area that's circled in the figure map, and then the hot spot eventually disappears for a while. This can be called the "Fireman Way Hot Spot" for the name of a street that runs up the middle of it. The dates July 8 through August 20, 2008, can be assigned for the time period.

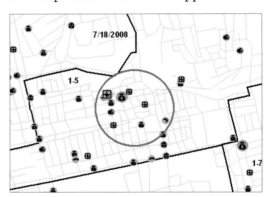

2 **Close Animation Manager.**

Create hot spot annotation layer

Now, you'll use the Drawing tool to draw the circle shown in the figure from the preceding exercise, and then convert that encircled area, which represents an emerging hot spot, to a feature layer for annotation.

1 From the Menu bar, click Customize > Toolbars > Draw.

2 On the Draw toolbar, click the Draw Shapes arrow and select Circle.

3 Use the previous figure for guidance to click on the map in the middle of the red circle and draw a radius of approximately the same size by dragging the Drawing tool away from the point. Then release it.

4 Right-click the circle and select Properties. Change the fill color to No Color, the outline color to Mars Red, and the outline width to 2. Click OK. You can adjust the location and size of the circle, if necessary, by dragging its handles.

5   Click the circle to select it; then click the Drawing arrow and select Convert Graphics To Features. Go to the Output shapefile box, click the Browse button, and go to your chapter 6 folder in MyExercises. Change the Save As type to shapefile, change Name to **FiremanWay.shp**, and click Save. Select the "Automatically delete graphics after conversion" check box. Then click OK > Yes.

6   In the table of contents, click List By Drawing Order, and then drag FiremanWay to the top.

Add attribute value to an annotation layer

To finish off the annotation layer, you need to add a value for labeling to the attribute table. The layer has a single feature, the circle, and its single record forms the attribute table.

1   From the Menu bar, click Customize > Toolbars > Editor.

2   On the Editor toolbar, click Editor > Start Editing and click FiremanWay. Click OK.

3   In the table of contents, right-click FiremanWay and select Open Attribute Table.

4   Click in the Name cell, type **Fireman Way Hot Spot**, and press ENTER.

5   Click Editor > Stop Editing. Click Yes. Close the attribute table and close the Editor toolbar.

6   In the table of contents, right-click FiremanWay and select Label Features.

7   Turn Fireman Way off. The animation turns this layer on when needed.

Add group layer for annotation to the animation

This time, instead of clearing the existing animation, you can add another track to it—a group layer that can be set to turn a portion of the animation on and off. This will take a few steps, starting with default values, and then narrowing the time points for turning elements on and off.

1   On the Animation toolbar, click the Animation arrow and select Create Group Animation. Click OK.

2   Open Animation Manager and click the Tracks tab. Be careful not to remove Time Animation track 1, and remove all *group* animations except for the first one, FiremanWay, by selecting group animation rows to be removed, and then clicking Remove. If you remove Time Animation track 1 by mistake, you'll have to clear the animation and re-create the time and group tracks.

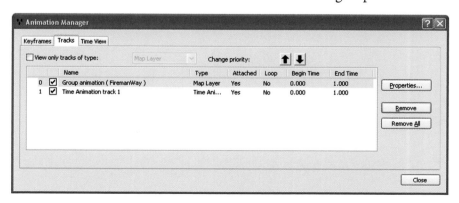

3   Click the Time View tab and drag the vertical red slider to the right until you get to 7/22/2008.

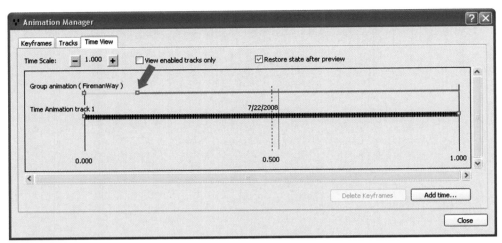

4   Drag the right green square of the Group animation (FiremanWay) line to the right to 7/22/2008.

5   Drag the red slider to the left to 7/8/2008 and drag the left green square of the group line to 7/8/2008.

6   Drag the red slider anywhere to the left of 7/8/2008.

7   Click to clear the "Restore state after preview" check box. Animation of group layers is sensitive to the end state. If you have the red circle turned on, it will start on but will end turned off, but if you start with the red circle turned off, it will turn on and off as desired.

8   In Time View, drag the red slider to the right of the second green square on the Group animation line. This step turns the red circle on between the two green squares and turns it back off to the right of them. For the rest of this computer session, the group layer will function as desired.

9   Close Animation Manager.

Play pin map animation

1   Open the Animation Controls toolbar.

2   Change Frame Duration to **1** second and type **170** in the From frame and **220** in the To frame.

3   In the table of contents, turn FiremanWay on.

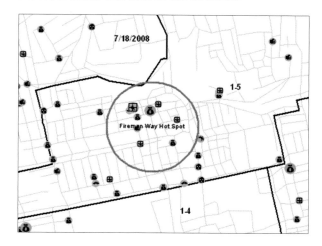

4   Play the animation. Now, you and other viewers can clearly see the emergence, persistence, and extinguishing of the hot spot over the course of about two weeks. The date stamp and hot spot label flicker when played in ArcMap, but that will stop when you create a movie file in the next Your Turn assignment.

5   Close the Animation Controls toolbar.

Save pin map animation

1   On the Animation toolbar, click the Animation arrow and select Save Animation File.

2   In the Save Animation window, go to your chapter 6 folder in MyExercises and change File Name to **FiremanWayHotSpot.ama**. Click Save. Now, you will always be able to load the animation in the future and play it when needed.

YOUR TURN   **YOUR TURN**   YOUR TURN   YOUR TURN   YOUR

Create a video file called **FiremanWay.avi** and save it to your chapter 6 folder in MyExercises. It will take a few minutes for ArcMap to create the file. When it is ready, play the recording.

Assignment 6-1

Build an animation for hot spots of illicit drug dealing

Drug dealing hot spots tend to be small, composed of only a half dozen blocks or so.

For this assignment, analyze 911 calls for illicit drug dealing by using an animation that's similar to the one for the field officers' pin map. Using a group layer, include circles and annotation to identify hot spots.

Create new map document

Create a new map document called **Assignment6-1YourName.mxd** and save it to your assignment 6-1 folder in MyAssignments. Use relative paths for your map document. Add the following map layers and data:

+ Streets, Rivers, and PoliceSectors from the Pittsburgh geodatabase in the Data folder.

+ Two copies of CADDrugsSummer08 from the Police geodatabase in the Data folder. The attribute EntryDate is the date of occurrence.

Symbolize layers

+ Symbolize Streets, Rivers, and PoliceSectors using best practices, including labels where relevant.

+ Symbolize one copy of CADDrugsSummer08 as the temporal context incidents for a four-week period controlled by the animation. Symbolize the other copy as the current day's incidents. For example, use point markers similar to those in the ATM robberies animation.

Build animations

There are two identified hot spots to animate. Each hot spot has the name of a street or intersection that is included in it, although the hot spots also include drug dealing points from nearby streets. The two significant streets are (1) Wilson Avenue and (2) the 800 block of Suismon Street. Use the Select By Attributes tool and the FENAME attribute of the Streets layer to find the streets and, thus, the hot spots. Use the following steps:

1 Zoom in on a hot spot and use a 1:5,000 scale with ArcMap at full screen. Draw a circle around the cluster of 11 drug calls in the vicinity of Wilson Avenue and Cutler Street for the Wilson Avenue hot spot. Convert the circle to a shapefile saved to your assignment 6-1 folder. Edit a value for Name and a label for annotation.

2 Create a bookmark for the hot spot.

3   Optional task: Create an XY file called **DateStampWilsonHotSpot.cvs,** including the attributes X, Y, and Date, that has coordinates for a point above the hot spot and dates for June 1 through August 31, 2008. Give this file time properties in ArcMap and do the same for your two drug call map layers.

4   Create an animation for the Wilson Avenue hot spot with two tracks, one a group animation and the other a time animation, as in tutorial 6-2. Save the animation as **WilsonHotSpot.ama** to your assignment 6-1 folder.

5   Select the 11 drug calls in the hot spot area and study the corresponding attribute table records to pick start and end dates for the hot spot. Set your animation to start running two weeks before the emerging hot spot starts and two weeks after it ends if there is enough data to do so or until your data runs out. **Hint:** To get frame numbers for the needed date interval, use the attribute table of the DateStampWilsonHotSpot.csv Events layer, if you have it. Sort by date and select records to get counts that correspond to the frame numbers by using the start and end dates of the interval.

6   Repeat these steps to create the Suismon Street animation.

What to turn in

Use a compression program to compress and save your assignment 6-1 folder to **Assignment6-1YourName.zip**. Turn in your compressed file.

Assignment 6-2

Build an animation for leading indicators of serious crime

Certain less serious crimes tend to occur before serious violent crimes in an emerging hot spot. If such leading behavior is present and detectable, police can then be proactive and work to prevent serious violent crimes.

For this assignment, use animations to help identify hot spots based on leading indicators.

Create new map document

Create a new map document called **Assignment6-2YourName.mxd** and save it to your assignment 6-2 folder in MyAssignments. Use relative paths for your map document. Add the following map layers and data:

* Spatial context layers: Streets, Rivers, and Tracts from the Pittsburgh geodatabase in the Data folder.

* Three copies of Offenses2008 from your chapter 6 folder in MyExercises. If you did not have EndDate = DateOccur + 28 in Offenses2008, create it now.

Symbolize layers

1   Symbolize Streets, Rivers, and Tracts using good practices, including labels where relevant.

2   Zoom in on the area of Pittsburgh east of its point, where the three rivers merge, including tracts 42003050100 and 42003050900 (see the figure map).

3   Symbolize and label the three copies of Offenses2008 as shown in the figure map.

4   Using the Draw tool, create the circle shown in the figure map, and convert it to a shapefile, saving it to your assignment 6-2 folder.

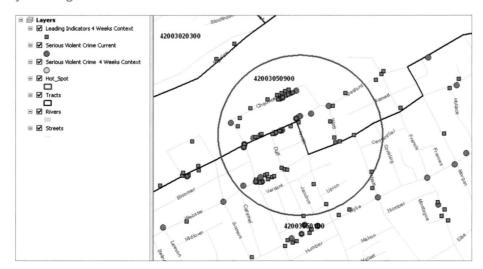

Create definition queries

You cannot include dates in definition queries of layers you plan to animate. The animation toolbar controls the date. You can create definition queries for other attributes, including crime types. Create the following definition queries:

- **Leading Indicators 4 Weeks Context**—include leading- indicator crime types of Simple Assault (hierarchy 10), Drug Violations (hierarchy 18), Public Drunkenness (hierarchy 23), and Disorderly Conduct (hierarchy 24). Use OR connectors.

- **Serious Violent Crimes Current** and **Serious Violent Crimes 4 Weeks Context**—include hierarchy 1 crimes–4 (Murder-Manslaughter, Forcible Rape, Robbery, and Aggravated Assault). Again, use OR connectors for hierarchy values.

Create, run, and save animation

Include the three crime point layers in the animation. When you run the animation, do so for just the summer months of June–August, corresponding to frames 153–244.

Save your animation file as **Assignment6-2YourName.ama** to your assignment 6-2 folder.

Analyze animation

Study your animation within the circled area. Does it appear as though the leading indicators occurred before any serious violent crimes?

What to turn in

Use a compression program to compress and save your assignment 6-2 folder to **Assignment6-2YourName.zip**. Turn in your compressed file.

OBJECTIVES

Identify spatial clusters
Use kernel density smoothing
Conduct Getis-Ord Gi* test for hot spot analysis

Chapter 7

Conducting hot spot analysis

Hot spot analysis is a well-accepted and widely used component of crime analysis. The objective is to automatically identify crime hot spots on a map. ArcMap has tools for this task that are available in ArcToolbox—in both the Spatial Statistics and Spatial Analyst tools. In this chapter, you learn how to put some key components of these tools to work.

Hot spot analysis concepts

Several studies have shown that very small parts of urban areas, hot spots, generate a very large number of crimes (for example, see Weisburd et al. 2004). Hot spots tend to persist, so identified hot spots make good targets for law enforcement and crime prevention. The objective of hot spot analysis is to automatically identify and estimate the boundaries of crime hot spots—spatial clusters of crime incidents for a certain period of time—to provide a "snapshot" of crime. Hot spots are also dynamic—they emerge, persist, decline, and re-emerge. The field officers' pin map in chapter 3 and corresponding animations in chapter 6 help identify such dynamic behavior because they include crime map layers from different time frames, providing a time context. You can see newer crimes in the context of recent but older crimes.

Spatial clustering and surface estimation are two approaches to automatically identifying hot spots. Spatial clustering involves searching for crime incidents that occur unusually close to each other. Surface estimation assumes there is an expected or average crime density (crimes per unit area) distributed over space that is smooth and continuous. The peaks of this surface define the hot spots. In this chapter, you learn how to use both clustering and smoothing techniques.

Spatial clustering

Some approaches to hot spot analysis use statistical methods, in particular, spatial statistics. As is the case with other advanced topics in this book, this introduction provides a brief description of underlying theories and methods and refers you to other resources for more in-depth coverage. You can find excellent documentation with additional references for the ArcToolbox Spatial Statistics toolbox in ArcMap by visiting ArcGIS Desktop Help.

Internet keyword search　　　**spatial statistics**

One simple test you use in this chapter is the Average Nearest Neighbor index. This Spatial Statistics tool can be used to test for spatial clustering. The index uses the ratio of the average distance between crime locations and their nearest-neighbor crime locations for a data sample that is divided by the same measure but for a random rearrangement of crime locations. If the ratio or index is significantly less than 1, it is evidence of spatial clustering. It is expected that crime would be clustered; for instance, many parts of an urban area have low or no crime rates (such as cemeteries, steep hillsides, and bodies of water) while other areas have high crime rates (such as public housing projects, run-down commercial areas, and gang territories). A test such as the Average Nearest Neighbor index is designed more to test how much clustering there is than to test whether or not clustering exists.

The Spatial Statistics toolbox also provides more sophisticated tests for clustering based on spatial autocorrelation, a measure of spatial clustering that uses aggregate data, such as the number of drug calls for service per block. Two such tests are Moran's I and Getis-Ord Gi* (pronounced G-i-star). Moran's I statistic, created by Patrick Moran, is a measure of spatial autocorrelation, or the degree to which a variable at a location has values similar to the same variable at nearby locations. The Getis-Ord Gi* test, named for its creators, Arthur Getis and Keith Ord, finds the locations and extents of point clusters in two-dimensional space. You will be using the Getis-Ord Gi* test for hot spot analysis in this chapter. A single block may have a high crime count, but it is not considered a hot spot unless there are other nearby blocks that also have high crime counts (that is, there is significant positive spatial autocorrelation).

Surface estimation using kernel density smoothing

Surface elevation is commonly used with rasters to illustrate crime density through surface estimation. A raster map is similar to a photograph and consists of a rectangular matrix of very small square grid cells, each with its own solid color. Raster maps can be used to portray surfaces that can then be used to depict varying levels of crime. You will create raster maps in this chapter that represent crime density—the expected or mean level of crime per unit area—through the use of surface elevation. The higher the expected crime count per unit area, the higher the surface peak. But rather than representing mountain peaks, the elevation in your raster maps will represent crime density peaks, or crime hot spots.

ArcGIS Desktop Help   **"What is raster data?"**

Kernel density smoothing is a leading method of surface estimation. "Kernel" is a mathematical term for a kind of function that can be used for averaging. Kernel density smoothing assigns a bell-shaped surface (the kernel) over every crime location, with the highest point of the bell directly over the crime location. Then at each point on the map, the kernel density smoothing method simply sums up all the bell surfaces overhead. Nearby crimes contribute a lot to the sum while distant crimes add only a small amount. The more crimes clustered near each other, the higher the surface at that point. The result, the kernel density surface, changes smoothly over space. For example, in the figure, the clustered drug calls for service result in crime density peaks, shown in red.

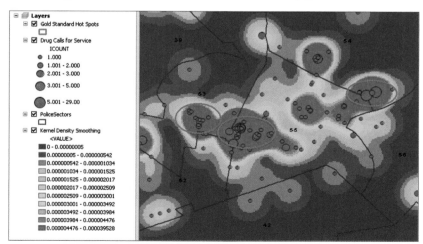

This map shows drug calls for service for a three-month period, gold standard hot spots, and a kernel density smoothing surface.

ArcGIS Desktop Help   **kernel density smoothing**

Hot spot analysis with raster maps amounts to estimating the crime surface over a jurisdiction, and then selecting a threshold elevation to yield a decision on what constitutes a hot spot. Any area (collection of raster cells) that is at or above the threshold is then considered a crime hot spot. The threshold in the figure map is the minimum crime density that corresponds to the boundary of the red areas. The red areas have crime density at the threshold or higher.

Calibrating kernel density smoothing for hot spots

The critical parameter for calibrating (or tuning) kernel density smoothing is the search radius (or radius of the bell surface placed over each crime). A relatively long radius results in "long rolling hills" for the surface, whereas a short radius results in sharp peaks. Of course, the search radius you need depends on the nature of the hot spots you want to identify.

Implementing hot spot methodology in practice, however, including kernel density smoothing, has led to a great deal of confusion, largely because there has been no "gold standard" approach to designating true hot spots for use in calibrating hot spot estimation methods. Some crime analysts use "rules of thumb" based on expert opinion for kernel density calibration, but there is no way to determine how such rules perform without having a target in place.

The approach used in this chapter to arrive at these gold standards, or assumed true hot spots, is to have crime analysts study crime pin maps and use their expert judgment to manually draw boundaries around the areas they perceive as hot spots. All that is needed is a good sample of expert hot spots; you do not have to exhaustively identify all hot spots in a dataset. Three gold standard hot spots are shown in the kernel density smoothing figure.

In kernel density smoothing, a search radius must be found that generates a crime density surface in which the peaks approximate the gold standard hot spot boundaries. This is not difficult, as you will see in the following exercises. The red areas of the kernel density smoothing surface in the kernel density smoothing figure do a reasonable job of approximating the gold standards identified there. In effect, this approach automates crime analysts' expert judgment by calibrating hot spot methods to replicate crime analysts' expertise as much as possible. The result is a hot spot method that saves time, is comprehensive, and provides consistent hot spots for police work. See "Empirical Calibration of Time Series Monitoring Methods Using Receiver Operating Characteristic Curves" (Cohen, Garman, and Gorr 2009) for an in-depth approach to calibrating methods for detecting the emergence of crime hot spots.

Conducting hot spot modeling

The workflow in the following exercises starts with testing for the existence of spatial clusters, moves on to kernel density estimation of hot spots calibrated by using expert judgment, and finishes with confirmation of the kernel density results by using the Getis-Ord Gi* hot spot analysis. You'll be using the tools that ArcMap provides in ArcToolbox, which simplifies the work. All you have to do is open a tool's parameter input form, select the inputs, and then run the tool to get the outputs.

Tutorial 7-1

Testing for crime spatial clusters

The null hypothesis (or possibility that there is no significant clustering) is a random distribution of points. If crimes are spread out uniformly and randomly, with no spatial clustering, a good strategy would be to use randomized patrols. If, however, crimes are spatially clustered, a strategy of targeted patrols in clustered or hot spot areas makes sense. The nearest-neighbor test that you'll run next makes a determination of spatial clustering versus the uniformly random, null hypothesis.

Open map document

1   Open Tutorial7-1.mxd in ArcMap. The starting map document opens with 911 drug calls for summer 2008 (June through August). You can see what appears to be quite a bit of spatial clustering for drug calls.

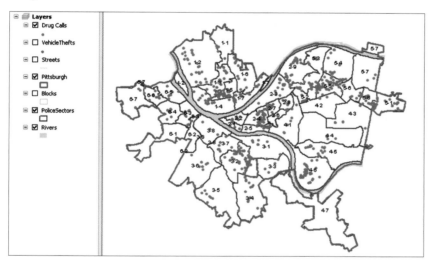

2   Save your map to your chapter 7 folder in MyExercises.

Create size-graduated point markers

To make clustering look even more apparent on the map, you can start by counting the number of drug calls at each address, and then use size-graduated point markers to symbolize Drug Calls. The larger the point symbol, the more drug calls at the same location. Many drug-dealing locations have multiple calls for service.

1   On the Standard toolbar, click ArcToolbox.

2   In the ArcToolbox window, expand the Spatial Statistics toolbox and then the Utilities toolset. Then double-click the Collect Events tool.

3   For Input Incident Features, select Drug Calls. For Output Weighted Point Feature Class, click the Browse button, go to the chapter 7 geodatabase in MyExercises, and type **DrugsCollected** for Name. Click OK > Close. Leave the ArcToolbox window open.

4   In the table of contents, right-click DrugsCollected, select Properties, and click the Symbology tab. Then click the Classify button and select Quantile for Classification Method. Click OK. Click the Template button and select Mars Red for color. Click OK twice. Now, with size-graduated point markers, it is even more evident that there is a lot of clustering. Try zooming in on a few police sectors that have clusters for a closer look. Turn Streets on while zoomed in. Next, you'll examine whether the nearest-neighbor statistical test provides further confirmation.

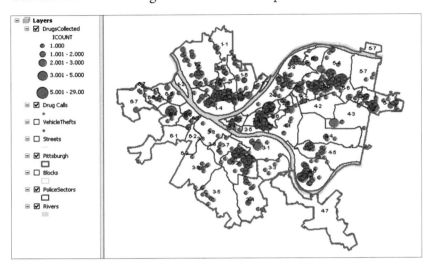

Test for spatial clustering

1   In the Spatial Statistics toolbox, expand the Analyzing Patterns toolset and double-click the Average Nearest Neighbor tool. The tool needs the total area of the study region (Pittsburgh). If you were to open the Pittsburgh layer's attribute table, you would see that its area is 1,626,651,299 sq ft. Note that to get a more realistic test, you could subtract the areas of steep hillsides, water areas, cemeteries, and other land types or uses where you'd expect no drug dealing. Then results would be less statistically significant, however, because the null hypothesis would have higher crime densities per the area reduction. That is, crime clusters would be less dense in relation to the null hypothesis.

2   Enter selections as shown in the figure.

3   Click OK.

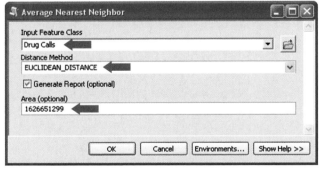

4   Click in the message window that appears after the tool runs and see what's inside. You should get a warning message that there were bad records that the tool ignored, which is fine.

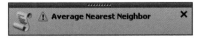

These are simply records that did not get geocoded from police records to points on the map, because they had invalid street addresses or the street centerline map was incomplete or inaccurate. Scroll down in the Results window where you'll also get the following output of the test:

```
Average Nearest Neighbor Summary
Observed Mean Distance: 249.565425
Expected Mean Distance: 630.183834
Nearest Neighbor Ratio: 0.396020
z-score: -36.974592
p-value: 0.000000
```

The test results provide evidence that drug calls are heavily clustered in Pittsburgh because the observed mean distance to the nearest neighbors of drug calls is significantly less than the expected mean distance under the null hypothesis that drug calls are uniformly and randomly distributed. The important statistic is the z-score, a common measure used in statistics that is calculated by subtracting the estimated mean of the distribution at hand from a data observation, and then dividing the difference by the sampled standard deviation. If the z-score has a value that is less than -2 or -3, it is evidence that the clustering does not occur by chance alone. In this case, the z-score is -36.97, which indicates a high level of statistical significance to the clustering of the crime data. In other words, there is just a small possibility that the clustering happened by chance alone and is therefore a phenomenon with some causal basis.

5 **Close the Results window.**

YOUR TURN **YOUR TURN** YOUR TURN YOUR TURN YOUR

Turn Vehicle Thefts on and turn Drug Calls and DrugsCollected off. Do you think vehicle thefts are spatially clustered? It would be easier to answer this question by visual inspection if vehicle thefts had size-graduated point markers, so create **VehicleTheftsCollected** and save to the chapter 7 geodatabase in MyExercises. Test for spatial clustering by using the Average Nearest Neighbor statistic with Vehicle Thefts as the input. You should get a z-score of -13.4, which is highly significant but less so than the same test for drug calls.

Visually scan for drug-dealing clusters

In preparation for using smoothing methods to estimate hot spots, you can try exercising your judgment about hot spots. The objective in this exercise is to identify hot spots for directing patrols, so you'll want hot spots that are fairly small. You can draw ellipses around areas that you believe are drug-dealing hot spots. You will have to zoom in to see potential hot spots.

1 **Turn DrugsCollected on, turn any other point layers off, and turn the Pittsburgh layer off.**

2 **Zoom in on the eastern portion of Pittsburgh and turn Streets on.**

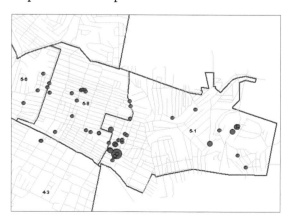

3   From the Menu bar, click Customize > Toolbars > Draw.

4   On the Draw toolbar, click the Draw Shapes arrow and select Ellipse. Then draw an ellipse roughly as shown in the figure, adjusting the ellipse's position, shape, and size by dragging its handles.

5   Right-click the ellipse, select Properties, and click the Symbol tab. Select No Color for the fill color, a dark blue for the outline color, and 2 for the outline width. Click OK.

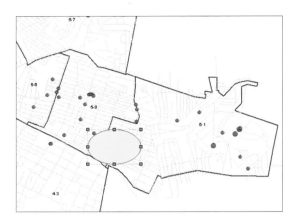

N YOUR TURN   **YOUR TURN**   YOUR TURN   YOUR TURN   YOU

Zoom in on other parts of Pittsburgh, look for two more hot spots, and draw ellipses around them. Make sure no graphics are selected, right-click the Layers label at the top of the table of contents, and select Convert Graphics To Features. Then click the Browse button for "Output shapefile or feature class," select "File and Personal Geodatabase feature classes" from the "Save as type" menu, and save **MyExpertHotSpots** to the chapter 7 geodatabase in MyExercises. Select the "Automatically delete graphics after conversion" check box. Click OK > Yes.

Add sample of supplied hot spot ellipses

Now, you know how to create expert-drawn hot spots. To save time and prepare for the following exercises, you'll next add a layer that has a larger sample of expert hot spot ellipses.

1   Click Add Data, go to the chapter 7 geodatabase in FinishedExercises, and add SuppliedHotSpots to the map.

2   Symbolize the added layer with No Color fill and a size 2 green outline.

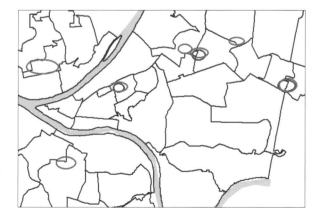

3   Compare your ellipses with the supplied hot spot ellipses. The three supplied ellipses with dark blue outlines represent the work you've done so far, and those with green outlines are the ones supplied for use in the following exercises. Perhaps you have ellipses that overlap the supplied ellipses. Neither selection is exhaustive of all potential hot spots, but you do not need to have all hot spots identified—just a good sampling of them. The objective is to train the kernel density smoothing method to approximate your judgment or someone else's. So in the next tutorial, you'll learn how to calibrate kernel density smoothing to produce a drug call surface that matches the SuppliedHotSpots layer.

4   Save your map document.

Tutorial 7-2

Using kernel density smoothing

The Spatial Analyst extension makes it simple to estimate density surfaces. All you need to do is set up the processing environment, and then supply a few inputs to a tool.

Activate Spatial Analyst and make environment settings

Spatial Analyst is one of several ArcMap extensions that users can purchase to add more functionality to their maps. With Spatial Analyst installed on your computer, you must activate it to be able to use it. You'll also need to change some overall settings, called Environment Settings, that affect all the tools in ArcToolbox.

1   Open Tutorial7-2.mxd in ArcMap.

2   Save your map document to your chapter 7 folder in MyExercises.

3   From the Menu bar, click Customize > Extensions. Select the Spatial Analyst check box to turn the extension on. Click Close. There are no apparent changes, but now you can use Spatial Analyst.

4   If necessary, click ArcToolbox on the Standard toolbar to open the ArcToolbox window.

5   In the ArcToolbox window, right-click the ArcToolbox icon and select Environments. Lengthen the Environment Settings window so you can see all the possible settings.

6   Click Workspace and make selections as shown in the figure.

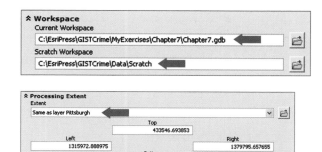

7   Click Workspace to close the panel. Then click Processing Extent and select "Same as layer Pittsburgh." Selecting this layer supplies the coordinates of the bounding rectangle, or extent, of any new layer that you create by using Spatial Analyst.

8   Click Processing Extent to close the panel. Then click Raster Analysis to expand that panel and enter selections as shown in the figure. When finished, close the Raster Analysis panel. A small cell size such as the default "1" increases the

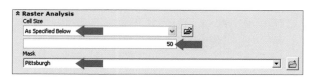

smoothness of the appearance of the raster map when zoomed in, but it also markedly increases computation times and the requirements for hard-disk storage. Setting the cell size to 50 ft to a side is adequate. Raster maps are all rectangular in shape. Specifying a mask limits the display of such a map to a jurisdiction boundary, such as Pittsburgh. ArcMap assigns all raster cells outside the mask No Color, thus making other portions of the map invisible.

9   Click OK to save settings and close the Environment Settings window.

Apply kernel density smoothing

The critical parameter to vary in the kernel density method is the search radius. The search radius includes points within that distance of a grid cell used to estimate the expected, or "smoothed," drug calls per square foot. The search radius will be relatively small because the expert cluster ellipses are small—around 2,000 ft along the long axis and 1,000 ft on the short axis. So, try 3,000, 1,000, and 500 ft for the search radius and see if one of these radii generates peak areas that match the green ellipses.

1   In ArcToolbox, expand the Spatial Analyst toolbox and then the Density toolset. Then double-click the Kernel Density tool.

2   Enter selections as shown in the figure.

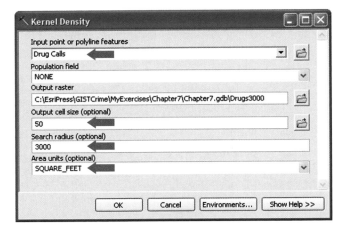

3   Click OK. This is the smoothest of the three surfaces you'll create, because it has the largest radius. In fact, it looks too smooth, as if it could use more peaks and valleys.

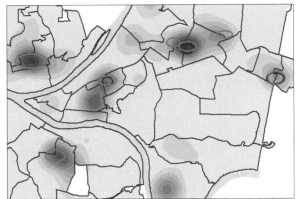

Symbolize density surface

Step 1 avoids a potential problem in symbolizing the raster map. You need to execute this step just once for a given map document.

1   From the Menu bar, click Customize > ArcMap Options. Click the Raster tab and then the Raster Catalog tab. Type **1000000** for "Maximum number of rasters for color matching." Then click OK.

2   Double-click Drugs3000 in the table of contents and click the Symbology tab.

3   Click Classify.

4   For Classification Method, select Standard Deviation, and for Classes, select 1/3 Std Dev. Click OK.

5   Select the color ramp that ranges from red to yellow to blue.

6 Click the Symbol heading above the resulting color chips in the Symbology panel and select Flip Colors so that low densities are blue and high densities are red.

7 Click OK.

8 Turn the DrugsCollected layer on and off for comparison with the kernel density and supplied hot spots.

9 Try zooming in on hot spots. This level of smoothing, with a search radius equal to 3,000 ft, does a nice job of capturing large hot

3,000 ft search radius.

spots in the red-colored raster cells. It might serve well as a map for the public—say, with the use of neighborhoods instead of police sectors. You'll need smaller hot spots, however, to direct police patrols. Drug dealing and gang turfs are more on the order of several blocks, which is quite a bit smaller than these areas.

YOUR TURN **YOUR TURN** YOUR TURN YOUR TURN YOUR

Run kernel density smoothing two more times, once with a search radius of **1000** ft, called **Drugs1000**, and once with a search radius of **500** ft, called **Drugs500**. Save the new raster layers to the chapter 7 geodatabase in MyExercises. Symbolize these layers as you did for the 3,000 ft layer. The 500 ft radius appears to be too short, resulting in smaller hot spots than identified by the expert ellipses, while the 1,000 ft radius does a fair job of matching them. Zoom in on several hot spots in the 1,000 ft radius density map and see how well the density smoothing does at this scale. The results are reasonable and useful. If you have new data for drug calls in Pittsburgh, you can use smoothing with a 1,000 ft search radius and be comfortable that Spatial Analyst will automatically find all the hot spots for you. When you are finished investigating the various density maps for hot spots, save your map document.

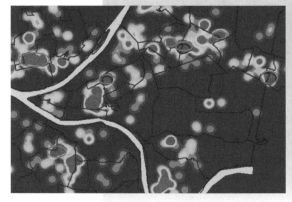

1,000 ft search radius.

500 ft search radius.

Create contour polygons defining a drug hot spot

The blue to red color ramp you have been using has hot colors that correspond to the threshold levels needed to define hot spots. Part of the reason this choice of color ramps works is because you used standard deviations for break points in the color ramp, and the high standard deviations pertain to the very high values of the distribution. Next, you'll learn how to convert the boundaries of the hot spot cells of the 1,000 ft radius map into vector contour lines stored as polygons. The polygon format makes the results easier to use and display.

1   In the ArcToolbox window, expand the Spatial Analyst toolbox and then the Surface toolset. Then double-click the Contour List tool.

2   Enter selections as shown in the figure.

Hint: Type the value **0.000004784** in the text box directly under the Contour Values label, and then press the Add button to get the value entered as shown in the figure.

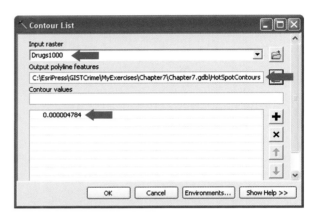

The contour value of 0.000004784 corresponds to the start of the second-highest range (reddish orange color chip) in the 1,000 ft search radius density map. So, the resulting contours will enclose hot spots that are reddish orange and red on the color ramp.

3   Click OK.

4   Turn all density maps off and turn SuppliedHotSpots on.

5   Symbolize the resulting HotSpotContour layer with a red outline of width 2. The result is a simple representation of drug hot spots that can be used for directing patrols. Notice that the contours from kernel density smoothing fit the supplied expert hot spots reasonably well. Also, some hot spots cross the boundaries of more than one police sector, suggesting that corresponding patrol units should coordinate patrols in the affected

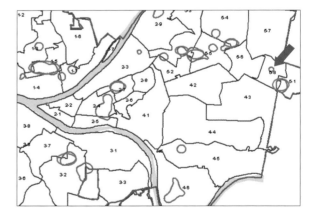

areas. Finally, kernel density smoothing found more hot spots than there were in the set of expert hot spots, but the latter dataset was intended only as a representative sample for training kernel density smoothing and not as the definitive set of hot spots.

6   Save your map document.

Edit hot spot contours

Suppose upon visual examination you feel that some of the areas defined by contours are too small or otherwise not drug hot spots. In that case, you can simply edit the contour map and delete unwanted polygons.

1   In the table of contents, click List By Selection and make HotSpotContour the only selectable layer.

2   From the Menu bar, click Customize > Toolbars > Editor.

3   On the Editor toolbar, click Editor > Start Editing, Select HotSpotContour1000. Click OK.

Next, you'll delete all polygons that are roughly the size of the contours indicated in HotSpotContour within Police Sector 5-3.

4   On the Standard toolbar, click the Select Features tool  and select a polygon you wish to delete by drawing a rectangle around it.

5   Open the HotSpotContour attribute table, right-click the row selector of the highlighted row (gray cell on the left side of the record), and click Delete Selected. Both the record and the polygon disappear.

YOUR TURN YOUR TURN YOUR TURN YOUR TURN YOUR

Use the Select Features tool, pressing SHIFT to select one or two more small contour areas, and then delete them in the attribute table. When you are finished, click Editor > Stop Editing. Click Yes. Save your map document.

Tutorial 7-3

Conducting Getis-Ord Gi* test for hot spot analysis

The Getis-Ord Gi* test identifies clusters of points that have higher values than expected by chance. The test, which you learn how to use in the following exercises, requires data aggregated from individual crime point locations to crime counts for small areas represented by centroid points. Aggregation is required because the test expects there to be variations in data, so, for example, the data cannot consist of all 1's. If an area's crime count is high and its neighbors' crime counts are also high, the Getis-Ord Gi* test concludes that these areas are part of a hot spot. A block is a good-size area to use for much of crime analysis, including this test, because the hot spots used for directing patrols or other police interventions need to be for small areas, and crime problems related to the hot spots are often at the scale of blocks. Creating the needed aggregate data takes several steps.

Open map document

1   Open Tutorial7-3.mxd in ArcMap.

2   In the table of contents, click List By Drawing Order.

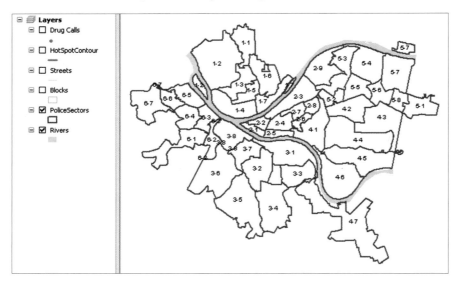

3   Save your map document to your chapter 7 folder in MyExercises.

Add XY centroid coordinates to a polygon attribute table

The centroid of a polygon is the point from which the polygon would balance if it were cut from cardboard and placed on a pencil point. (The case of a quarter-moon-shaped polygon is an exception because its centroid is not inside its boundary.) So, a centroid is a kind of center of a polygon that makes a handy map layer for displaying certain kinds of polygon information. In this exercise, you add centroid coordinates to the Blocks attribute table. You need to do this task just once to have the results available for future use.

1   In the table of contents, right-click Blocks and select Open Attribute Table.

2   Click Table Options > Add Field. Type **X** for Name and select Double for Type. Click OK.

3   Repeat step 2 to create the **Y** field.

4   Right-click the X column heading and select Calculate Geometry. Click Yes. Select X Coordinate of Centroid and click OK.

5   Repeat step 4 to compute the Y Coordinate of Centroid.

OBJECTID *	Shape *	BLKIDFP00	Shape_Length	Shape_Area	X	Y
1	Polygon	420032507001029	946.199601	41055.393258	1337035.571578	418516.163244
2	Polygon	420032107002003	1627.614639	88197.0056	1335744.758819	417237.087807
3	Polygon	420032507001033	924.436852	42089.51039	1336748.908627	418315.409402
4	Polygon	420032507001030	550.128088	18246.792107	1337290.261691	418562.555745
5	Polygon	420032107002012	1100.189362	63867.763186	1336630.172077	417079.809194
6	Polygon	420032107002046	1405.710049	78135.671335	1335996.603714	415873.945166

6   Click Table Options > Export. Click the Browse button and go to the chapter 7 geodatabase in MyExercises. Select "File and Personal Geodatabase tables" from the "Save as type" menu, change Name to **BlockCentroidsTable,** and click Save. Click OK > No.

7   Close the table.

Create block centroid map layer

Creating a feature layer for block center points, or block centroids, is another one-time task. The results can be used over the life of the crime mapping system or until the block map is revised.

1   From the Menu bar, click Windows > Catalog.

2   Expand the Catalog tree to the contents of the chapter 7 geodatabase in MyExercises.

3   Right-click BlockCentroidsTable and click Create Feature Class > From XY Table.

4   Click the Coordinate System of Input Coordinates button, click Import, and select Blocks. Click Add > OK.

5   Change Output Feature Class to **BlockCentroids** in the chapter 7 geodatabase. Click Save > OK.

6   Drag BlockCentroids from Catalog to the table of contents under Drug Calls.

7   Close the Catalog window. Now you have a feature layer for block centroids. In the next exercise, you'll count the drug calls by block and attach the counts to the block centroids layer you just created.

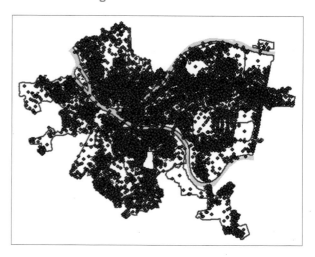

Overlay points with polygons

Every time you have new data for hot spot analysis, you'll need to perform this task and the remaining ones in this section. In this exercise, you use spatial overlay to assign a block ID to each crime point. Note that spatial overlay arbitrarily assigns crime points located at a street intersection to a single block, and thus does not double-count these points. There is no "correct answer" for assigning such points to blocks, so this is the best solution. Crime points with street addresses have a 20 ft offset from street centerlines, and thus fall within the correct blocks.

1 In ArcToolbox, expand the Analysis toolbox and then the Overlay toolset. Then double-click the Spatial Join tool.

2 Enter selections as shown in the figure.

3 Click OK. Now every drug call record in the DrugsBlocks layer includes the block ID, BLKIDFP00, for the block that it's in.

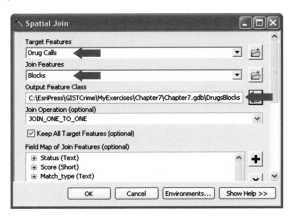

Aggregate drug calls by block

1 In the table of contents, right-click DrugsBlocks and select Open Attribute Table.

2 Right-click the BLKIDFP00 column heading and select Summarize.

3 For item 3 of the Summarize dialog box, specify Output Table as **DrugsAggregated** and save to the chapter 7 geodatabase in MyExercises. Click OK > Yes. The resulting table has the number of drug calls per block in Pittsburgh. The table in the figure shows rows 25–30.

OBJECTID *	BLKIDFP00	Count_BLKIDFP00
25	420030203001086	1
26	420030305001010	2
27	420030305002001	2
28	420030305002002	13
29	420030305002005	14
30	420030305002007	3

4 Close the DrugsAggregated table.

Join drug calls by block to a block centroids layer

1 In the table of contents, right-click BlockCentroids. Then click Joins and Relates > Join.

2 Make selections as shown in the figure.

3 Click OK > Yes.

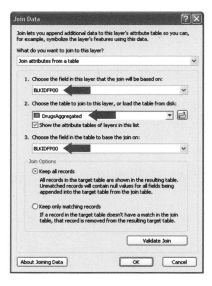

Export feature layer with joined data to a new feature layer

If you open the BlockCentroids attribute table, you can see that there are many blocks with no drug calls and that the joined count attribute, Count_BLKIDFP00, has "NULL" for many values. You need to replace these nulls with a zero (0), but ArcMap will not let you calculate joined fields. The remedy is to export BlockCentroids to a new feature layer. That way, ArcMap includes the joined attributes as permanent attributes in the resulting feature layer, and you can replace the nulls with zeros.

1   Remove DrugsBlocks from the table of contents. This version of DrugsBlocks has individual drug call points. You can now replace this layer with the version that has drug counts per block, presented as block centroids.

2   In the table of contents, right-click BlockCentroids. Then click Data > Export Data.

3   In the Export Data dialog box under Output Feature Classes, click the Browse button and go to the chapter 7 geodatabase in MyExercises. Type **DrugsBlocksAggregated** for Name, select "File and Personal Geodatabase feature class" from the "Save as type" menu, and click Save. Click OK > Yes.

4   Right-click DrugsBlocksAggregated and select Open Attribute Table.

5   Click Table Options > Select By Attributes.

6   In the top panel, or attributes panel, double-click "Cnt_BLKIDFP00", click the Is button, and then click Get Unique Values. In the values panel, double-click NULL to yield `"Cnt _ BLKIDFP00" Is NULL`. Click Apply.

7   Right-click the Cnt_BLKIDFP00 column heading in the attribute table and select Field Calculator.

8   In the Field Calculator dialog box, type **0** in the bottom panel and click OK.

9   Close the attribute table.

10 From the Menu bar, click Selection > Clear Selected Features. You have now cleared the blocks with no drug calls from the map.

11 Save your map document.

Conduct Getis-Ord Gi* test

In this exercise, you can see how the hot spot analysis from kernel density smoothing compares to a sophisticated statistical approach, the Getis-Ord Gi* test. The key idea here is to match the search radius of both methods by using a 1,000 ft fixed band (search radius) for Getis-Ord Gi*, which is the same search radius chosen for kernel density smoothing. As with most statistical tests, the Getis-Ord Gi* test focuses on false positive rates. A false positive rate in this case is the fraction of blocks that are not true members of hot spots yet are falsely classified as hot spot areas by the Getis-Ord Gi* test. By tradition, statistical testing most often uses conservative false positive rates of 0.01 or 0.05 to establish decision rules for classifying potential hot spots.

1   In the ArcToolbox window, expand the Spatial Statistics toolbox and then the Mapping Clusters toolset. Then double-click the Hot Spot Analysis (Getis-Ord Gi*) tool.

2 **Enter selections as shown in the figure.** The selection of a fixed band of 1,000 ft corresponds to the best-case kernel density map, which similarly weighs neighboring points by use of a 1,000 ft search radius.

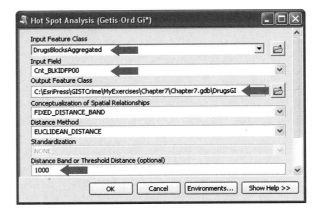

3 **Click OK.**

4 **After ArcMap adds the new DrugsGI layer to the table of contents, turn the HotSpotContour layer on and zoom in on several of the areas that have red and orange point markers.** The red and orange point markers indicate blocks that are significantly hotter than other blocks and, as you can see, correspond reasonably well to the HotSpotContour boundaries. The orange point markers correspond to a false positive error rate of 0.05 (z-value = 1.65 and higher) and the red point markers correspond to a 0.01 false positive error rate. This test result provides confirmation that the hot spots you found from kernel density smoothing are statistically significant. Next, you'll try a more lenient false positive error rate of 0.10 (z-value = 1.29 and higher) for orange point markers, which is closer to police preferences. Police organizations are willing to spend resources on a certain amount of false positives in order to find real positives, or hot spots.

5 **In the table of contents, double-click DrugsGI and click the Symbology tab.**

6 **Click the Range -1.649999–1.65000 (yellow point marker), type 1.29, and press the TAB key.** Even though you type just one value, 1.29, ArcMap fills out the entire range and completes a revised set of categories. This makes the orange category correspond to a

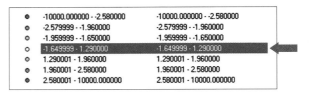

false positive rate of 0.10, so orange, orange-red, and red categories are now considered significant.

7   **Click OK.** There is some expansion of the number of blocks that are determined to be in hot spots. The Getis-Ord Gi* analysis considers the circular hot spot that was derived from the smoothing method, indicated in the figure, to be a false positive, whereas all the other smoothing hot spots are confirmed. Overall, the Getis-Ord Gi* test provides confirmation of using the judgmental approach to calibrate kernel density smoothing.

8   **Save your map document and exit ArcMap.**

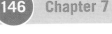

Assess impact of date duration on hot spots

For some policing purposes, the more recent the information, the better. So, the question becomes, how short can the time interval be (year, quarter, month, week) for hot spot analysis to be meaningful? Working in the opposite direction, the more that reliability of statistical methods and modeling (such as smoothing) is improved, the larger the sample size. So, you might conclude that for a given crime, there is a minimum threshold at which it makes no sense to attempt hot spot analysis short of that time interval.

For this assignment, investigate the impact of time intervals by using the drug call data applied to the exercises in this chapter. The hot spots for three summer months looked fine in the exercises, so try hot spots for a month and then a week to see how short a time interval is feasible for crime analysis.

Create new map document

Create a new map document called **Assignment7-1YourName.mxd** and a new file geodatabase called **Assignment7-1.gdb** and save them to your assignment 7-1 folder in MyAssignments. Include the following map layers in your map document:

+ Streets, Rivers, Pittsburgh, and PoliceSectors from the Pittsburgh geodatabase in the Data folder
+ CADDrugsSummer08 from the Police geodatabase in the Data folder and DrugsCollected from the chapter 7 geodatabase in FinishedExercises

Save all new map layers that you create to your assignment 7-1 geodatabase.

Symbolize layers

+ Symbolize Streets, Rivers, Pittsburgh, and Tracts, using best practices.
+ In the table of contents, rename DrugsCollected to **DrugsCollectedSummer08**. Symbolize this layer, using graduated point markers with a Circle 2 point marker in Mars Red and the quantile classification method with five classes. Save the results as a layer file named **DrugsCollected.lyr** to your assignment 7-1 folder for reuse when classifying collected drugs for other time periods.

Create definition queries

From CADDrugsSummer08, make two additional layers: **CADDrugsAugust08**, which has drug calls for August 2008, and **CADDrugsLastWeekAugust08**, which has drug calls for the last seven days of August 2008.

Conduct spatial analysis

Create a Word document called **Assignment7-1YourName.docx** and save it to your assignment 7-1 folder. Run the nearest-neighbor test to see if each kind of crime is spatially clustered and include the results in your document.

Create a kernel density map for each time duration—summer, month, and week. Use Pittsburgh as the mask, 50 ft for the grid cell size, and 1,500 ft for the search radius. Render these maps using the symbology used in this chapter: standard deviations with one-third increments and the blue to red color ramp.

Hint: You can save a lot of time by using collected offense points as the inputs to density estimation, instead of aggregating points to blocks or other polygons. Although it may be better to aggregate points to small polygons because that is a smoothing step in itself, collected points nonetheless provide a reasonable approximation for quick analysis. So, to do this, use the Collect Events tool, which is in the Spatial Statistics toolbox in the Utilities toolset. You will need to use this tool for each queried time period: summer, month, and week. Use your exported layer file, DrugsCollected, to keep the symbology consistent.

Zoom in on the North Side of Pittsburgh, the area north of the three rivers. Export each density map with the corresponding collected-drugs layers displayed. Add the maps to your Word document and label each with a caption such as "Figure 1. Density Map for Drug Calls for Service: Pittsburgh North Side, June–August 2008."

Comment on each map's hot spots. Consider only the hottest density interval, red. Besides the density map for the summer period, are the density maps for the two shorter periods useful? Explain. Does the nearest-neighbor test provide any guidelines?

What to turn in

Use a compression program to compress and save your assignment 7-1 folder to **Assignment7-1YourName.zip**. Turn in your compressed file.

Assignment 7-2

Examine broken-windows theory for simple and aggravated assaults

Broken-windows theory predicts that soft crimes will harden over time to serious crimes. Simple assault is a major leading indicator of serious violent crimes, and aggravated assault is the serious violent crime in this crime category. A simple assault is an intentional act by a person that causes apprehension in another of imminent harm or contact. It is an attempt to frighten another person. Aggravated assault is more serious and includes intent to do serious bodily harm—that is, to rob, rape, or kill. If simple assaults tend to harden into aggravated assaults, you'd expect for hot spots for simple assaults to contain or overlap hot spots for aggravated assaults. While not a complete test of the broken-windows theory, this expectation of connected hot spots is nevertheless predicted by the theory.

For this assignment, analyze these two types of crime, simple assaults and aggravated assaults, to see if there is a correlation between them.

Create new map document

Create a new map document called **Assignment7-2YourName.mxd** and a new file geodatabase called **Assignment7-2.gdb** and save them to your assignment 7-2 folder in MyAssignments. Add the following map layers:

+ Streets, Rivers, Pittsburgh, and PoliceSectors from the Pittsburgh geodatabase in the Data folder
+ Two copies of Offenses2008 from the Police geodatabase in the Data folder

Save all new map layers to your assignment 7-2 geodatabase.

Symbolize layers

+ Symbolize Streets, Rivers, Pittsburgh, and Tracts using best practices.
+ Turn both copies of Offenses2008 off in the table of contents.

Create definition queries

Limit each copy of Offenses2008 to June through August 2008. Add to your date query to make one copy **Simple Assaults: Summer 2008** and the other **Aggravated Assaults: Summer 2008**.

Conduct spatial analysis

Create a Word document called **Assignment7-2YourName.docx** and save it to your assignment 7-2 folder. Run the nearest-neighbor test to see if each kind of crime is spatially clustered and include the results in your document.

Create a kernel density map for each crime type, using Pittsburgh as the mask, 50 ft as the grid cell size, and 1,500 ft as the search radius. Render these maps using the symbology used in this chapter: standard deviations with one-third increments and the blue to red color ramp.

Hint: Use the same hint as in assignment 7-1: instead of aggregating points to blocks or other polygons, use the collected points as the input to kernel density smoothing.

Create a contour map for each density map using the start value of the highest-density class (red). Symbolize each contour map using a different line color and turn off all map layers except the contours and spatial context layers. Export the map (or better, a layout) as **Contours.jpg** to your assignment 7-2 folder and add it to your Word document. Comment on the evidence for broken windows in regard to simple and aggravated assaults. Is there a tendency for hot spots for simple assaults and hot spots for aggravated assaults to overlap?

What to turn in

Use a compression program to compress and save your assignment 7-2 folder to
Assignment7-2YourName.zip. Turn in your compressed file.

References

Cohen, J., S. Garman, and W. L. Gorr. 2009. Empirical calibration of time series monitoring methods using receiver operating characteristic curves. *International Journal of Forecasting* 25:48–61.

Weisburd, D., S. Bushway, C. Lum, and S. Yang. 2004. Trajectories of crime at places: A longitudinal study of street segments in the City of Seattle. *Criminology* 42:283–321.

Part 3
Building a crime mapping
and analysis system

OBJECTIVES

Download and preprocess basemaps and spatial data
Extract jurisdiction maps from basemaps
Create new map layers from basemaps
Digitize features by using basemaps as guides

Chapter 8

Assembling jurisdiction maps

A police department can build its jurisdiction maps from basemap layers that are available on the Internet for free download. In this chapter, you learn how to (1) download, clean up, and import basemaps and data; (2) extract map features as subsets of basemaps for a jurisdiction such as a city; (3) merge basemaps to produce task force regions that are larger than a single basemap; and (4) create new map layers from base layers by dissolving interior lines to create new polygons and by digitizing custom car beats and block watches from tracing lines in basemaps. All these skills are essential in building a new crime mapping and analysis system. They are also useful in revising existing systems as new map layers become available and in updating existing map layers.

Basemaps and jurisdiction maps

Many countries have developed a digital map infrastructure that provides basemap layers for political boundaries, statistical boundaries, physical features, railroads, street centerlines, and other features. In the United States, features as detailed as block-long street centerlines are publicly available as basemaps from the U.S. Geological Survey and the U.S. Census Bureau, while individual property boundaries and building footprints are generally produced by local governments. Basemaps are also available from vendors, who improve the accuracy, completeness, and detail of basemaps. The federal basemaps you use in this chapter are Census Bureau TIGER (Topologically Integrated Geographic Encoding and Referencing) maps, designed for census taking but widely used for other purposes.

Internet keyword search **TIGER maps**

Essential basemaps for crime mapping

Many types of basemaps are needed to create crime maps. Essential maps available by county and used in this book include the following:

+ *County subdivision boundaries*, which have boundaries for municipalities, many of which have their own police departments.

+ *Census tract boundaries*, which subdivide counties into statistical areas of approximately 4,000 population each. Census tracts are neighborhoods with roughly homogeneous socioeconomic characteristics, which makes them useful for construction of police sectors consisting of one or more contiguous census tracts. You can also use census tract maps to display census data on demographics, including poverty and youth populations.

+ *Street centerlines*, which are block-long street segments that include house numbers for the starting and ending addresses on both sides of the street. Streets form the basis for mapping crime incident data as points through the process of geocoding or address matching (see chapter 9). The majority of census tract boundaries are street centerlines.

The accompanying figure illustrates these maps, as downloaded from the Census Bureau Web site for Allegheny County, Pennsylvania. The police jurisdiction of interest is the City of Pittsburgh in the center, shown with a bluish green boundary. There are 129 other municipalities in the county, 416 census tracts, and 81,646 block-long street segments.

Much of crime mapping can be carried out by starting with just these three basemaps. Additional basemaps can be used to provide information on the spatial context of crime. Examples include schools, convenience stores, parks, topography, and aerial photos.

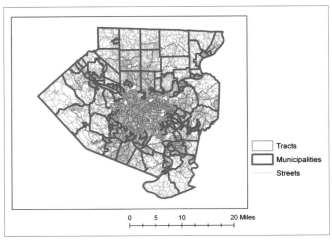

Basemaps for Allegheny County, Pennsylvania.

Extraction of jurisdiction maps

The first step in building a crime mapping and analysis system is to obtain needed basemaps from the Internet or other sources. Often, the boundary of a police jurisdiction is a subset of a basemap; for example, Pittsburgh is one polygon in the Allegheny County, Pennsylvania, basemap of county subdivisions. The next step is to extract the polygon(s) that constitute the jurisdiction from the basemap. The jurisdiction boundary, once obtained, becomes the "cookie cutter" used to extract other features such as census tract boundaries and streets from a basemap. A map for a multijurisdictional task force may require merging adjacent basemaps to create the study area.

Many organizations, including police departments, have administrative boundaries that are analogous to the private sector's sales territories. Police departments often have police sectors (car beats) that have an assigned patrol unit for every shift. For example, Pittsburgh has 42 car beats that are each made up of an average of three of its 132 census tracts, with a patrol unit for every shift. Given a table that lists each census tract and the car beat that is assigned to it, you can create a map of car beat boundaries from census tracts by "dissolving" the interior boundary lines of census tracts with the use of the Dissolve tool in ArcMap. Alternatively, if it is desirable to have custom car beats that do not follow census tract boundaries, you can use streets and other features from basemaps as guides, and then digitize the custom areas by using your pointer to trace them on the map.

Spatially enabled aggregate data

Mapping aggregate data such as population or monthly crime counts to census tracts, car beats, or other areas within a jurisdiction can prove useful to crime investigators. In this chapter, for example, you download a census table that lists the number of persons living in poverty per census tract. You then join that data table to a census tract map in ArcMap by using the unique location identifiers that are included in both the data and the map—namely, the unique tract number the U.S. Census Bureau gives to each census tract. Once the data table is joined to the map, you can use the population data to color-code census tracts to show levels of poverty. Thus, the tract number becomes the geocode (or attribute that spatially enables the data), allowing the color-coded census tracts to be displayed on the map, revealing patterns of poverty.

Assembling police jurisdiction map layers

In the exercises that follow in this chapter, you will be assembling police jurisdiction map layers for the Penn Hills Police Department in Allegheny County. The reason you are working on map layers for Penn Hills is because the map layers for the Pittsburgh Police Bureau are already complete and available for use in this book's data files.

Tutorial 8-1

Downloading and preprocessing basemaps

The first task for assembling a jurisdiction map is simply to download free basemaps and census data. You will need to process some of the downloaded files before using them, but those instructions follow in subsequent exercises in this chapter.

Download TIGER basemaps from the Census Bureau Web site

If you have difficulty downloading files, you can find them in the Downloads folder in chapter 8 of FinishedExercises and use those files for processing in later exercises in this chapter.

1   Start a Web browser such as Internet Explorer.

2   Go to `http://www.census.gov`, click the TIGER link under Geography, click the 2009 TIGER/ Line Shapefiles Main Page link under Previous TIGER/Line Shapefiles, and finally click "Download the 2009 TIGER/Line Shapefiles now". A shapefile is a simple file format for map layers, which was designed by Esri and is in wide use. Note that the wording and placement of the Web links cited may vary over time as newer versions of the files become available, but it is the 2009 TIGER/Line shapefiles that you need to download for the exercises that follow.

3   For state- and county-based shapefiles, select Pennsylvania. Click Submit.

4   For Pennsylvania county file, select Allegheny County. Click Submit.

5   Select files as shown in the following figure.

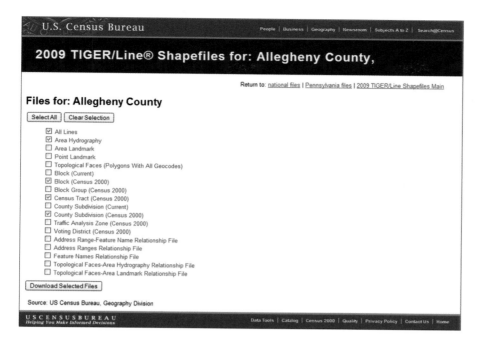

6   **Click Download Selected Files > Open. Extract the file to the Downloads folder in chapter 8 of MyExercises.** The result is five compressed files in a ZIP folder called 42_PENNSYLVANIA\42003_Allegheny_County:

- ◆ tl_2009_42003_areawater.zip
- ◆ tl_2009_42003_cousub00.zip
- ◆ tl_2009_42003_edges.zip
- ◆ tl_2009_42003_tabblock00.zip
- ◆ tl_2009_42003_tract00.zip

7   **Open a My Computer window and extract each of the compressed files to the Downloads folder.**

8   **Delete the 42_PENNSYLVANIA folder that has the compressed files to save space on your hard drive.** If you look in a My Computer window, you can see that each shapefile consists of five data files. For example, tl_2009_42003_areawater.shp is the shapefile for water features—hence, the .shp file extension. The shapefile stores the graphic features; the database file (.dbf) stores the attribute table; the shapefile projection (.prj) stores metadata on the coordinate system or projection of the map layer; and the shapefile index (.shx) contains an index to speed up processing of the shapefile. Note that "42" in the file names is the Census Bureau code for Pennsylvania and "003" is the code for Allegheny County.

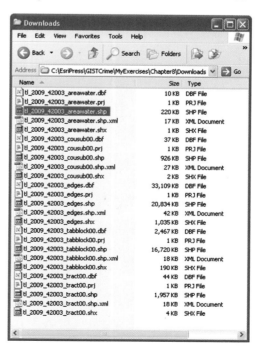

Download census data from the Census Bureau Web site

1   Go to the American FactFinder Web site at `http://factfinder.census.gov`.

2   In the left panel, click the Download Center button.

3   Click the Census 2000 Summary File 3 (SF 3) - Sample Data link. SF 3 data is from the long-form census, administered to a random subset of households, and then adjusted to reflect the entire population. The short-form census (SF 1 data) is administered to the total population and has basic population, age, race, ethnicity, gender, and family relationship data, whereas the long form adds data on educational attainment, income, employment, nature and place of work, nature of residence structure, and so on.

4   Under County Level, click All Census Tracts in a County (140), select the state of Pennsylvania, and select Allegheny County.

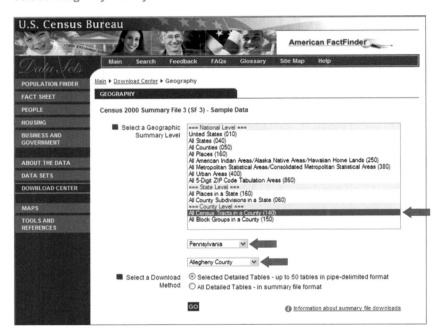

5   Click Go.

6   In the "Select one or more tables" panel, scroll down and click P87 Poverty Status in 1999 by Age. Then click the Add button. If you needed additional census data attributes, you could continue to add tables and the FactFinder wizard would build a single table with all the attributes of the input tables. Here, however, you need only the poverty data.

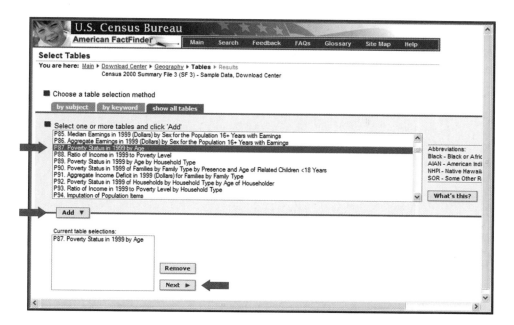

7   Click Next > Start Download > Open. Then extract each file to the Downloads folder. The result is three files, and the one with the needed census data is in dc_dec_2000_sf3_u_data1.txt.

8   Close your browser.

If you choose to skip the next optional exercise, you can find the results for later use in the chapter 8 folder in FinishedExercises.

Preprocess census data with the use of spreadsheet software (optional exercise)

The census data you just downloaded needs to be put in table form, with the top row containing attribute names and all remaining rows serving as raw data rows. Instead, the data now has *two* rows at the top that contain attribute information, and the rest of the rows serve as data rows. The easiest way to carry out this task so the data is in table form is to open the data in Microsoft Office Excel.

1   Start Excel.

2   Click the Office button and click Open.

3   In the Open window, change "Files of type" to All Files (*.*), browse through the Downloads folder in chapter 8 of MyExercises, and double-click dc_dec_2000_sf3_u_data1.txt.

4   In the Text Import Wizard, click the Delimited option, click Next, and click to clear the Tab option. Click Other and type the | character (press SHIFT+\). Click Next > Finish.

5   Press and drag your pointer in the column heading selectors from A through U to select those columns, and then double-click between any two selectors to resize column widths so that you can read all the cell contents.

6   Clear your selection. Then delete the following columns, identified in row 1, by right-clicking column selectors one by one, and then clicking Delete: GEO_ID, SUMLEVEL, GEO_NAME, and PO87010 through PO87017.

7   Rename row 1 attributes by replacing the values as follows:

GEO_ID2	**Tract**	P087005	**PopPov6T11**
P087001	**Pop**	P087006	**PopPov12T17**
P087002	**PopPov**	P087007	**PopPov18T64**
P087003	**PopPovU5**	P087008	**PopPov65T74**
P087004	**PopPov5**	P087009	**PopPov75U**

8   Right-click the row selector for row 2 and press DELETE.

9   Double-click the tab at the lower-left corner of the
     spreadsheet that has the label dc_dec_2000_sf3_udata1
     and type **AllCoPoverty** to give the tab a new label.

10   Name your spreadsheet **Poverty** and save it to your
     chapter 8 folder in MyExercises. Then close Excel.

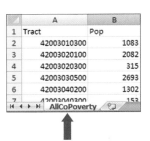

Project coordinates for the Census Bureau All Lines shapefile

All the shapefiles you downloaded from the Census Bureau TIGER site have geographical or spherical coordinates (latitude and longitude in decimal degrees) for location on a spherical representation of the world's surface. When plotted on a flat surface, such as your computer screen, these maps are distorted; for example, a midlatitude area, such as Allegheny County, has north- and south-oriented features that are shortened. ArcMap can project from spherical to ordinary rectangular coordinates for small areas of the United States or any other place in the world, resulting in maps that have very little distortion. While ArcMap can project on the fly every time you display a map layer, it is best to store maps in projected coordinates to save processing time on your computer. In this exercise, you project the All Lines shapefile you downloaded, tl_2009_42003_edges.shp, to state plane coordinates, a collection of projections covering all of the United States that is widely used by local governments.

1   Save a blank map document as **Tutorial8-1.mxd** to your chapter 8 folder in MyExercises.

2   Click File > Map Document Properties. Then click the "Store relative pathnames to data sources" check
     box. Click OK.

3   On the Standard toolbar, click Add Data. Then click the Browse button, go to the Downloads folder in
     chapter 8 of MyExercises, and click tl_2009_42003_edges.shp. Then click Add. If you zoom in on
     the center of the map, and then pause the pointer over the point where the three rivers meet, you
     can see that its coordinates are roughly -80° longitude (east-west) and 40° latitude (north-south).
     Next, you'll give the data frame called Layers in the table of contents a state plane projection.
     ArcMap will then transform any map layer in the data frame to the state plane coordinate system.

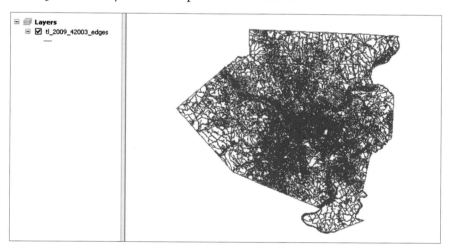

4   Right-click Layers at the top of the table of contents, select Properties, and click the Coordinate System tab.

5   In the "Select a coordinate system" panel, go to Predefined > Projected Coordinate Systems > State Plane > NAD 1983 (CORS96) (US Feet). Click NAD 1983 (CORS96) StatePlane Pennsylvania South FIPS3702 (USFt). Click OK > Yes. Now, the map layer looks as it would look from an airplane with correct north-south lengths. In the next exercise, you'll extract streets from this layer and save them permanently with state plane coordinates.

Extract streets from the Census Bureau All Lines shapefile

The tl_2009_42003_edges shapefile has all the lines that are used to build census blocks, including lines from streets, water features, and railroads. The smallest geographical area used by the Census Bureau is blocks, which are the "atoms" used to build all other geographical areas used for the tabulation of census data, including block groups, tracts, cities, and counties. In this exercise, you extract streets from the shapefile, using the FEATCAT (feature category) attribute for streets, to create a new feature class, AlleghenyCountyStreets. The FEATCAT code values are S for streets, R for railroad tracks, H for water (hydrology) features, and so on.

1   From the Menu bar, click Selection > Select By Attributes. Scroll down in the values panel list of attributes, double-click "FEATCAT", and click the = button. Click Get Unique Values and double-click 'S'. The resulting query criterion, "FEATCAT" = 'S', will select all features that are streets. "FEATCAT" = 'R' would select all railroad tracks, and "FEATCAT" = 'H' would select all hydrology (water) features.

2   Click OK.

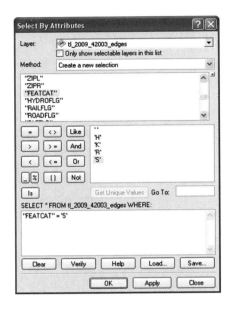

3   In the table of contents, right-click tl_2009_42003_edges and click Data > Export Data. Click the Browse button, go to your chapter 8 folder in MyExercises, change the Save As type to "File and Personal Geodatabase feature classes," double-click the chapter 8 geodatabase, and type **AlleghenyCountyStreets** for File Name. Click Save. Then click the Data Frame option. Click OK > Yes.

4   Remove tl_2009_42003_edges from the table of contents.

5   In the table of contents, right-click AlleghenyCountyStreets and select Open Attribute Table. You should have 90,587 block-long street segments for the county and many types of attributes. The MTFCC code is the type of street—for example, S1730 is an alley and S1400 is a city street. (For code meanings, see MAF/TIGER Feature Class Code [MTFCC] Definitions on the Census Bureau Web site by clicking the TIGER link, and then going to 2009 TIGER/line Shapefiles Main Page > Technical Documentation > Appendix F. Scroll down in the document to the codes that start with S.)

OBJECTID *	Shape *	STATEFP	COUNTYFP	TLID	TFIDL	TFIDR	MTFCC	FULLNAME
1	Polyline	42	003	51680896	224226873	210184971	S1730	Armour Way
2	Polyline	42	003	51637707	210180536	210188677	S1400	Stebbins Ave
3	Polyline	42	003	51680919	210195388	210180542	S1400	Shiras Ave
4	Polyline	42	003	51636863	210180557	210196060	S1400	Margaret St
5	Polyline	42	003	51637063	210180558	210188492	S1400	Giffin Ave
6	Polyline	42	003	51637077	210180564	210188485	S1400	Nobles Ln
7	Polyline	42	003	51637089	210180565	210188448	S1400	Trost Ave

6   Close the attribute table and save your map document.

Tutorial 8-2

Extracting jurisdiction maps

In this exercise, you extract Penn Hills layers from the Allegheny County basemaps you downloaded to create a jurisdiction map for Pen Hills. ArcMap has several useful tools and processes that make this work straightforward and simple.

Create new map document

The first step to building a map for the Penn Hills jurisdiction is to create a new map document, set its properties, add the county subdivision map you downloaded from the Census Bureau TIGER Web site, and set the map document data frame projection to the state plane coordinate system.

1   Save your map document as **Tutorial8-2.mxd** to your chapter 8 folder in MyExercises.

2   In the table of contents, turn AlleghenyCountyStreets off.

3   On the Standard toolbar, click Add Data. Browse through the Downloads folder in chapter 8 of MyExercises (or alternatively, through the Downloads folder in chapter 8 of FinishedExercises if you did not download map layers from the Census Bureau earlier in this chapter), and double-click tl_2009_42003_cousub00.shp. If you get a Geographic Coordinate Systems Warning message, click "Don't warn me again ever" and click Close. The map includes Penn Hills and two other municipalities you use in tutorial 8-4 to create a task force region.

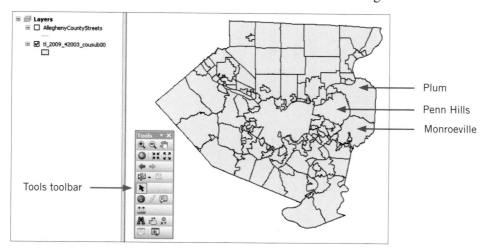

Extract "cookie cutter" feature class for Penn Hills

To make it easy to extract additional features from different map layers for the same jurisdiction, you can extract the shape of Penn Hills and create the Penn Hills cookie cutter. To do this, you'll use the Select Features tool to select Penn Hills, and then save the cookie cutter shape as a new map layer.

1   On the Tools toolbar, click the Select Features tool, and then click Penn Hills to select its polygon. For reference, see the figure from step 3 in the preceding exercise. The Penn Hills boundary turns the bluish green selection color, indicating that it is selected.

2   In the table of contents, right-click tl_2009_42003_cousub00. Then click Data > Export Data and select the data frame option.

3   Go to the chapter 8 geodatabase in MyExercises and type **PennHills** for File Name. Click Save > OK > Yes.

4   In the table of contents, right-click PennHills and select Zoom To Layer.

5   From the Menu bar, click Selection > Clear Selected Features.

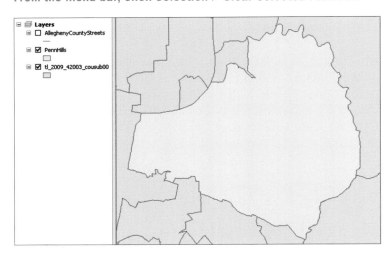

Create two additional feature classes consisting of outlines of municipalities in state plane coordinates, one called **Monroeville** for the municipality of Monroeville and the other called **Plum** for the municipality of Plum. Save them to the chapter 8 geodatabase in MyExercises. When you are finished, in the table of contents, right-click tl_2009_42003_cousub00, Monroeville, and Plum one at a time and click Remove. Clear any selections by clicking Selection > Clear Selected Features.

Symbolize map layers

1   Zoom in on Penn Hills.

2   Add the following map layers to your map document from the Downloads folder in chapter 8 of MyExercises:

- tl_2009_42003_areawater.shp

- tl_2009_42003_tabblock00.shp

- tl_2009_42003_tract00.shp

3   Symbolize map layers as follows:

- PennHills: hollow fill with a red outline in width 4

- tl_2009_42003_tract00: hollow fill with a black outline in width 2

- tl_2009_42003_tabblock00: hollow fill with a tan or brown outline in width 1

- tl_2009_42003_areawater: blue fill

You should be able to see the outlines of all polygon layers now, with the width 4 red outline for PennHills showing underneath the width 2 tracts and width 1 blocks.

4   At the top of the table of contents, click List By Drawing Order.

5   In the table of contents, drag layers so that their order from the top is PennHills, tl_2009_42003_areawater, tl_2009_42003_tract00, tl_2009_42003_tabblock00, and AlleghenyCountyStreets.

Clean up attributes from downloaded shapefiles

Now, you can delete some of the unwanted attributes from your downloaded shapefiles. This saves hard-disk storage space and makes it easier for users to find the attributes they need.

1 In the table of contents, right-click tl_2009_42003_tabblock00 and select Open Attribute Table.

2 Right-click the STATEFP00 column heading and click Delete Field. Click Yes.

3 Similarly, delete all other attributes except FID, Shape, and BLKIDFP00. ArcMap will not let you delete FID or Shape. BLKIDFP00 is the full identification value for blocks and is an identifier that is used throughout the country.

FID	Shape	BLKIDFP00
0	Polygon	420034271002001
1	Polygon	420034240001023
2	Polygon	420034240003023
3	Polygon	420034240001007
4	Polygon	420034240002006
5	Polygon	420034240003021

4 Close the attribute table.

YOUR TURN **YOUR TURN** YOUR TURN YOUR TURN YOU

Clean up the attribute table of tl_2009_42003_tract00: Delete all attributes except FID, Shape, and CTIDFP00. Also delete the following attributes from AlleghenyCountyStreets: STATEFP, COUNTYFP, TLID, TFIDL, TFIDR, SMID, and all attributes to the right of FEATCAT, including FEATCAT but excluding Shape_Length (which ArcMap will not let you delete). Close the table when you are finished and save your map document.

Use cookie cutter shape to extract map layers

Now, you are ready to extract additional features for Penn Hills.

1 From the Menu bar, click Selection > Select by Location and make selections in the Select by Location window as shown in the figure.

2 Click OK. Tracts that make up Penn Hills are all in the selection color, bluish green.

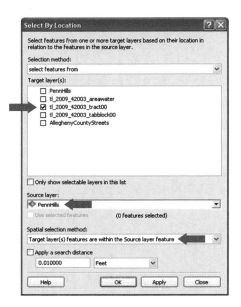

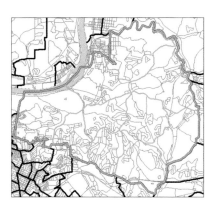

3 In the table of contents, right-click tl_2009_42003_tract00. Click Data > Export Data and click the data frame option. Go to the chapter 8 geodatabase in MyExercises and type **PennHillsTracts** for File Name. Click Save > OK > Yes.

4 From the Menu bar, click Selection > Clear Selected Features.

5 Symbolize the new layer with a hollow fill and black outline, width 2.

6 Right-click tl_2009_42003_tract00 and click Remove. Clicking Remove removes the layer from your map document but does not delete it from your files.

7 Save your map document.

YOUR TURN **YOUR TURN** YOUR TURN YOUR TURN YOUR

Extract blocks for Penn Hills from tl_2009_42003_tabblock00. Save the new feature class as **PennHillsBlocks** to the chapter 8 geodatabase in MyExercises. Remove tl_2009_42003_tabblock00 from the table of contents and symbolize PennHillsBlocks with a hollow fill and a medium gray outline, width 1.

Extract streets for Penn Hills from AlleghenyCountyStreets. Save the new feature class as **PennHillsStreets** to the chapter 8 geodatabase. Symbolize the new layer with a light gray line for streets (the default width is fine) and remove AlleghenyCountyStreets from the table of contents. Turn PennHillsStreets off for a moment so you can see PennHillsBlocks. Then turn PennHillsBlocks off and turn PennHillsStreets on. Save your map document.

Clip layer

The river boundary of Penn Hills should be included on crime maps because it is a natural barrier to social interactions with Pittsburgh, and that includes criminal interactions. So far, all the extractions you have done have been whole polygons or line segments. In this case, however, you need to extract only a portion of the river—the part that runs through Penn Hills—rather than the entire river polygon. The tool for this purpose is the Clip tool, which you can use with your Penn Hills cookie cutter to cut out only the portion of the river that is in Penn Hills. The Clip tool is in ArcToolbox.

1 On the Standard toolbar, click ArcToolbox.

2 In the ArcToolbox window, right-click the ArcToolbox icon and select Environments.

3 Click Output Coordinates. Under Output Coordinate System, select Same as Display. Click OK. In this environment setting, any new layers you create will have the same coordinate system as the Layers data frame in the table of contents—in this case, state plane.

4   In ArcToolbox, expand the Analysis toolbox and then the Extract toolset.

5   Double-click the Clip tool and enter selections as shown in the figure.

6   Click OK twice.

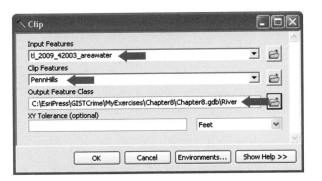

7   Remove tl_2009_42003_areawater from the table of contents.

8   Symbolize River with a blue fill. Close ArcToolbox.

9   Turn River, PennHillsBlocks, PennHillsTracts, and PennHills on and save your map document.

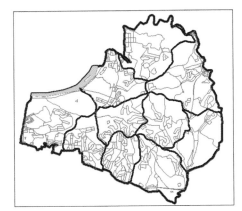

Tutorial 8-3

Joining census data to polygon maps

Because there are thousands of census variables, it would be impractical for the Census Bureau to provide all the variables in the attribute table of a polygon map layer. The files would be too large. So, you'll need to download a table that contains the census variables you want, and then join that table to the appropriate map layer attribute table.

Change data type of a geocode

The geocodes, or spatial identifiers, used to join two tables must match in both value *and* data type. However, the geocodes in census data tables are a text data type, while the same geocodes in polygon map layers are a numeric data type. So, you need to create a new attribute in one of the tables, keeping its existing values but changing the data type to match the other table.

1   Save your map document as **Tutorial8-3.mxd** to your chapter 8 folder in MyExercises.

2   On the Standard toolbar, click Add Data. Browse through the Downloads folder in chapter 8 of MyExercises and double-click Poverty.xls. Double-click AllCoPoverty$ to add this table to your map document. If you skipped the optional spreadsheet exercise using Excel earlier in this chapter, you can find the finished file in the Downloads folder in chapter 8 of FinishedExercises.

3 In the table of contents, right-click AllCoPoverty$ and click Open. Make the Tract column wider so you can see that it is right aligned (numeric), and then close the table. Numerals that are right-aligned are stored as numeric data, while numerals that are left-aligned are stored as text. This is true of most software packages, including ArcMap. So, attributes whose numeral values are labels and not intended for addition or subtraction, such as ZIP Codes, would normally be stored as text, while population values would be stored as numeric.

4 Right-click PennHillsTracts and select Open Attribute Table. Note that the geocode (CTIDFP00) is left-aligned (text).

5 Click Table Options > Add Field. Type **Tract** for Name and select Double for Type. Click OK.

6 Right-click the new Tract column heading and select Field Calculator.

7 In the Fields panel, double-click CTIDFP00 to add it to the Tract = panel at the bottom. Click OK. The new Tract column is set as numeric, and thus will join with AllCoPoverty$, whose tract values are also numeric.

OBJECTID ^	Shape ^	CTIDFP00	Shape_Length	Shape_Area	Tract
1	Polygon	42003523800	36752.396659	55517706.457839	42003523800
2	Polygon	42003523702	52307.878427	74961853.060558	42003523702
3	Polygon	42003523701	46057.84305	69251571.645435	42003523701
4	Polygon	42003523600	33794.695034	57522516.074251	42003523600
5	Polygon	42003523502	35943.448176	50624447.325147	42003523502
6	Polygon	42003523501	30594.455874	49351905.540206	42003523501

8 Close the table.

Join data table to a map layer

You can now join the census tract data table to the tract polygon map layer to provide census information for each tract in the map.

1 In the table of contents, right-click PennHillsTracts and click Joins and Relates > Join. Enter selections as shown in the figure.

2 Click OK.

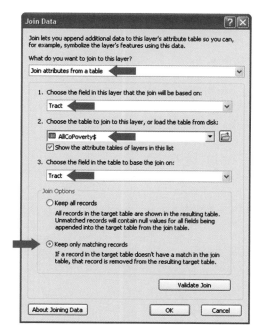

3   In the table of contents, right-click PennHillsTracts and select Open Attribute Table. Notice that each tract has poverty data joined to it. The null values correspond to tracts that had no poverty population in the corresponding age range. In the next exercise, you'll use the joined poverty data when you dissolve the tracts to create police sectors, or car beats, for Penn Hills. ArcGIS creates beat-level poverty data as a byproduct.

CTIDFP00	Shape_Len	Shape_Area	Tract	Tract	Pop	PopPov	PopPovU5	PopPov5	PopPov6T11
▶ 42003523800	36752.39665	55517706.45	4200352380	4200352380	5086	208	11	2	3
42003523702	52307.87842	74961853.06	4200352370	4200352370	5709	641	89	<Null>	85
42003523701	46057.84305	69251571.64	4200352370	4200352370	4547	485	46	38	49
42003523600	33794.69503	57522516.07	4200352360	4200352360	5276	234	14	<Null>	32
42003523502	35943.44817	50624447.32	4200352350	4200352350	1695	116	18	4	34
42003523501	30594.45587	49351905.54	4200352350	4200352350	5168	298	16	<Null>	15

4   Close the table and save your map document.

Tutorial 8-4

Creating new map layers from basemaps

There are several ways to create new map layers from existing basemaps. In this exercise, you learn how to (1) create centroids (a point map layer) from a polygon layer by using built-in ArcGIS functions and tools; (2) dissolve polygons to create larger aggregate polygons; and (3) merge adjacent polygon map layers to create a single layer.

Calculate polygon centroid coordinates for blocks

Given that you now have basemaps assembled, it is possible to create new maps as derivatives of the basemaps. The first map to create is for centroids of blocks. Centroids, or center points, have many potential uses in mapping. For example, you use block centroids in chapter 4 as a way of maintaining confidentiality in crime maps released to the public. Instead of showing the actual location of a crime incident, you add random numbers to the centroid coordinates, and thus plot the incident at a random location within the block polygon.

1   Save your map document as **Tutorial8-4.mxd** to your chapter 8 folder in MyExercises.

2   In the table of contents, turn all layers off, except PennHillsBlocks and PennHills.

3   Right-click PennHillsBlocks and select Open Attribute Table.

4   On the Standard toolbar, click Table Options > Add Field. Type **X** for Name and select Double for Type. Click OK.

5   Repeat step 4, except type **Y** for Name.

6   Right-click the X column heading and select Calculate Geometry. Click Yes. Select X Coordinate of Centroid for property. Click OK.

7   Repeat step 6, except select Y Coordinate of Centroid for property.

8   Leave the PennHillsBlocks table open.

Create centroid point feature class for blocks

With x- and y-coordinates computed for centroids, you can create a new map layer with just the centroid points.

1   Click Table Options > Export. For Output table, go to the chapter 8 geodatabase in MyExercises and type **PennHillsBlockCentroids** for Feature Class Name. Click Save > OK > Yes. Close the table.

2   On the Standard toolbar, click the Catalog window button 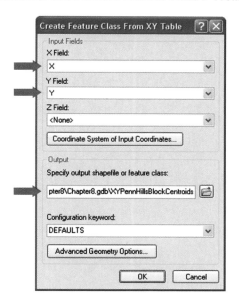, go to the chapter 8 geodatabase in MyExercises, and right-click PennHillsBlockCentroids. Click Create Feature Class > From XY Table.

3   Enter selections as shown in the figure.

4   Click OK.

5   Drag XYPennHillsBlockCentroids from the Catalog window to the top of the table of contents. Then close the Catalog window.

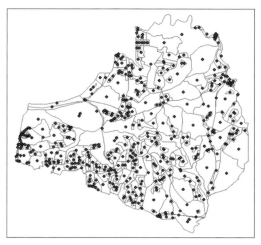

Create field for dissolving

Car beats can be created by combining adjacent census tracts, and then dissolving the interior lines of tracts to form new polygons. You can use the Dissolve tool in ArcToolbox to do this. First, you need to create a "dissolve" field in the census tract attribute table that records the car beat in which each tract is located. ArcMap then uses that field to dissolve the census tract boundaries to form the car beats. The map provided here shows the three desired car beats, with each car beat consisting of census tracts in the same color. In this exercise, you manually select all the tracts in a car beat to assign them the car beat number. With the dissolve field completed for all tracts, the Dissolve tool then has sufficient information to create the desired car beat polygons.

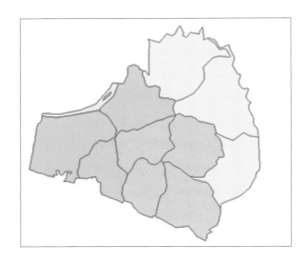

1   In the table of contents, turn PennHillsTracts on and turn all other layers off. Right-click PennHillsTracts and select Open Attribute Table.

2   Click Table Options > Add Field.

3   In the Add Field window, type **CarBeat** for Name, leaving Short Integer the default for Data Type. Click OK and leave the attribute table open.

4   On the Tools toolbar, click the Select Features tool, press SHIFT, and click in each of the three leftmost tracts making up a car beat.

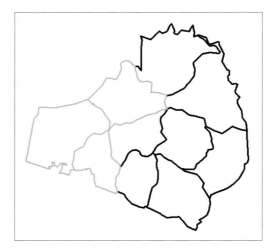

5   In the attribute table, scroll to the right, right-click the PennHillsTracts.CarBeat column heading, and select Field Calculator. The calculation you are about to perform applies only to the selected rows, which are all for car beat 1.

6   In the Field Calculator, click in the panel under PennHillsTracts.CarBeat = and type **1**. Click OK.

7   From the Menu bar, click Selection > Clear Selected Features.

YOUR TURN   **YOUR TURN**   YOUR TURN   YOUR TURN   YOUR

Repeat steps 4–6 to add the value 2 to the four blue car beats shown at the bottom center of the figure map at the start of this exercise. Then use the same process to add the value 3 to the three car beats to the right. Clear selected features. A sample of the finished attribute table is shown in the figure. When you are finished, close the attribute table.

PopPov18T64	PopPov65T74	PopPov75U	PennHillsTracts.CarBeat
115	28	32	1
342	39	40	3
248	18	35	3
125	28	23	2
32	13	9	3
218	26	<Null>	2

Dissolve polygons

With the work done of creating the CarBeat field for use in dissolving the related tracts, it is easy to create the car beat polygons.

1   Make sure no tracts are selected and close the attribute table, if it is still open.

2   Open the ArcToolbox window (if necessary, click ArcToolbox on the Standard toolbar).

3   Expand the Data Management toolbox and then the Generalization toolset. Then double-click the Dissolve tool.

4   Make selections as shown in the figure, but do not click OK yet.

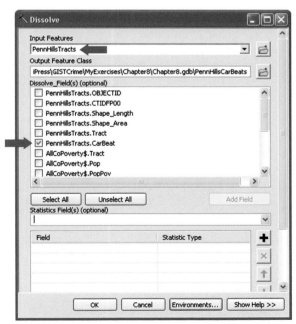

The following set of steps shows how to select tract data attributes to aggregate to the car beats and how to sum up the data from associated tracts.

5   Click the Statistics Field(s) arrow and select AllCoPoverty$.Pop.

6   Click in the Statistic Type field adjacent to the AllCountyPoverty$.Pop value you just selected and select SUM.

7 Repeat steps 5 and 6 for **AllCoPoverty$.PopPov.**

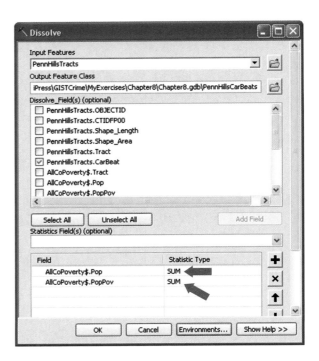

8 **Click OK and label the car beats by clicking the Labels tab in Layer Properties. Hint:** Make sure the "Label features in this layer" check box is selected.

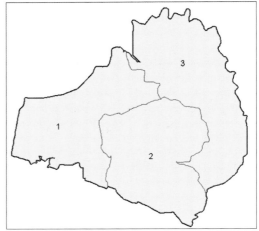

9 **In the table of contents, right-click PennHillsCarBeats and select Open Attribute Table.** The new car beat map layer has the three dissolved polygons representing the three car beats, plus aggregated census data for the total population and the population that's living in poverty.

OBJECTID ^	Shape ^	CarBeat	SUM_AllCoPoverty__Pop	SUM_AllCoPoverty__PopPov	Shape_Length	Shape_Area
1	Polygon	1	14518	1038	77832.478635	167388938.643512
2	Polygon	2	19890	1206	62707.898651	180030075.030719
3	Polygon	3	11951	1242	89390.081899	194837872.031147

10 **Close the table and save your map document.**

Assign aliases to attributes

The names that ArcGIS created for the attributes of aggregated variables are too long. While ArcGIS does not have the functionality to change the names of attributes, you can assign an alias to the names, so ArcGIS can use the shorter names you pick instead of the actual names.

1   Right-click PennHillsCarBeats, select Properties, and click the Fields tab.

2   In the left panel, click SUM_AllCoPoverty_Pop to select it, and in the Alias cell, replace the label by typing **Pop**.

3   Click the SUM_AllCoPoverty_PopPov label and type **PopPov** for Alias.

4   Turn off OBJECTID, Shape, Shape_Length, and Shape_Area.

5   Click OK.

6   Open the PennHillsCarBeats attribute table.

CarBeat	Pop	PopPov
3	11951	1242
2	19890	1206
1	14518	1038

7   Close the table.

Merge polygon map layers to create a new map layer

Earlier in this chapter, you extracted three new polygon layers from a Census Bureau basemap: PennHills, Monroeville, and Plum. For the sake of learning how to merge map layers, suppose you had downloaded these three map layers from a Web site and now need to merge them into a single map layer.

1   On the Standard toolbar, click Add Data. Browse through the chapter 8 geodatabase in MyExercises and click Monroeville. Click Add.

2   Repeat step 1 for Plum.

3   In the table of contents, turn layers on or off so that only PennHills, Monroeville, and Plum are on.

4   On the Tools toolbar, click Full Extent.

5   In the ArcToolbox window, expand the Data Management toolbox and then the General toolset. Then double-click the Merge tool.

6   Make selections as shown in the figure. **Note:** The attribute tables of map layers that are being merged have to match in both attribute names and data types, which is the case here.

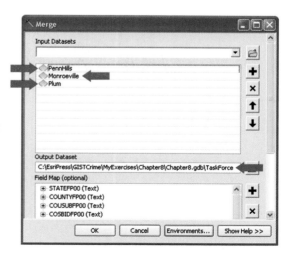

7 **Click OK.** Now, a single layer, the TaskForce feature class, has polygons for all three municipalities. Of course, if you wanted, you could continue to merge other layers, such as streets, for the task force region map layer.

8 Remove the Plum, Monroeville, and TaskForce map layers from the table of contents.

9 In the table of contents, right-click PennHills and select Zoom To Layer.

Tutorial 8-5

Digitizing features

You do not need any special equipment, such as a digitizing tablet, to do the digitizing in this exercise. Instead, you can use your pointer and basemap features as manuscript material to do "heads-up digitizing," in which you look up at your screen as a guide as you trace features with your pointer rather than looking down at a digitizing tablet. First, you'll digitize custom car beat polygons by using the Penn Hills boundary polygon and Penn Hills major streets. The ArcGIS trace editing function makes this task simple and fast, whereby you create a new empty feature class, and then add your own features. Second, you'll digitize neighborhood block watches as line features by using the snapping functionality in ArcMap to align newly digitized block watch lines with existing street lines.

Create empty feature classes

Sometimes, the police need to custom design car beats along lines that do not correspond to existing polygons. In these cases, you can use street centerlines and other lines to form new polygons into an original design. ArcMap has very sophisticated tools that allow you to piece together existing lines for this purpose. To use them, you first have to create an empty feature class in ArcCatalog to hold the new polygons. In this exercise, you create CarBeatsCustom to store the new custom car beats.

1 Save your map document as **Tutorial8-5.mxd** to your chapter 8 folder in MyExercises.

2 On the Standard toolbar, click the Catalog window button. In the Catalog tree, expand your chapter 8 folder in MyExercises if necessary so you can see the chapter 8 geodatabase icon.

3 Right-click the chapter 8 geodatabase and click New > Feature Class.

4 In the Create New Feature Class dialog box, type **CarBeatsCustom** for Name, select Polygon for Type, and click Next.

5 Click the Import button and select PennHills in the chapter 8 geodatabase. Then click Add and click Next three times.

6 Enter selections as shown in the figure to create two new attributes.

7 Click Finish.

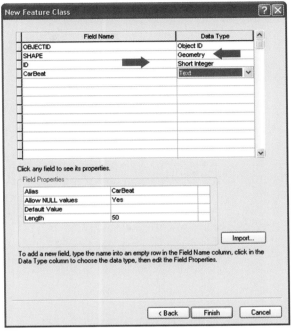

Create a new feature class called **BlockWatches** as a line feature map layer that uses the PennHills coordinate system. Create the attributes **ID** (short integer) and **Name** (text). Then add BlockWatches to your map document. Close the Catalog window when you are finished.

Set up map document for tracing

The Trace tool on the Editor toolbar can be used to create new custom car beats from streets and the Penn Hills boundary. The Trace tool creates an exact copy of whatever features you trace over with your pointer and places them in a new map layer. Suppose, for instance, that the Penn Hills Police Department wants to use the boundary of the municipality and its secondary or primary streets (excluding local or neighborhood streets) as potential car beat boundaries. Secondary or primary streets have CFCC (census feature class code) values of A35 or less in the TIGER PennHillsStreets layer. In this exercise, you learn how to create a new layer for these streets that can be used as a guide for digitizing.

1 In the table of contents, right-click PennHillsStreets and click Copy.

2 From the Menu bar, click Edit > Paste. You now have a second copy of PennHillsStreets in the table of contents.

3 In the table of contents, change the name of the new copy of PennHillsStreets to **PennHillsMajorStreets**.

4 Right-click PennHillsMajorStreets and select Properties. Click the Definition Query tab, and then click Query Builder.

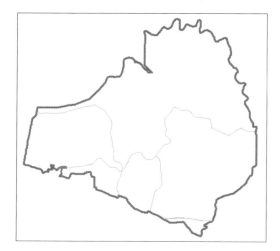

5 Build the query `"MTFCC" = 'S1100' OR "MTFCC" = 'S1200'`. Click OK twice.

6 Turn layers on or off in the table of contents so that only PennHillsMajorStreets, CarBeatsCustom, and PennHills layers are turned on.

In the next exercise, you'll digitize the selected streets and a portion of the Penn Hills boundary to form a car beat polygon. You will not need to make the selections shown in the figure map. They are there just so you can see the desired boundary of the new polygon.

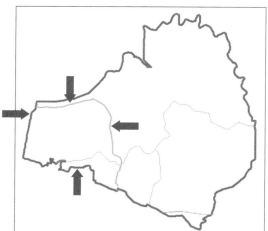

7 Use the Zoom In tool to zoom in on the lower-left corner of PennHills.

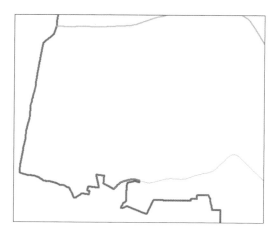

Digitize custom car beats

Now, you can use the Editor toolbar and the Trace tool to create new custom car beats from streets and the Penn Hills boundary. It can be challenging at first to use the Trace tool , but once you get the hang of it, it works very well.

1   From the Menu bar, click Customize > Toolbars > Editor.

2   On the Editor toolbar, click Editor > Start Editing. In the top panel, click CarBeatsCustom. Then click OK > Continue.

3   In the Create Features window, click CarBeatsCustom.

4   On the Editor toolbar, click the Trace tool button for polygons ⚃▾ (the fifth button to the right of the Editor arrow).

5   Click the lower-left corner of the PennHills map and trace along the boundary to the right (just move the mouse along smoothly without clicking). Keep the red PennHills boundary inside the head of the pointer. As you trace, you will see the traced line on the map. When you get to the right edge of the map in the next step, pan to the right. While pressing the letter C on your keyboard, pan in the direction you need to digitize and keep digitizing. You can also back up if needed while tracing. If you need to start over, double-click and press DELETE. If you lose the tracing pointer, click CarBeatsCustom in the Create Features window, and then click the Trace tool for polygons on the Editor toolbar.

6   When you come all the way around to the starting point, double-click to finish the polygon.

7   If you are satisfied with the resulting polygon, click Editor > Save Edits. If you wish to try again, click Editor > Stop Editing, do not save edits, and repeat steps 2–7.

Pan to the right in your map so you can see the middle of Penn Hills and the completed east boundary of your first polygon. Trace a second polygon roughly as shown in the figure map at left. Proceed by using the Editor toolbar to stop editing, and then start editing, following the preceding steps. Click the Trace tool and be sure that adjacent polygons share the same boundary. The second polygon in the left figure map has two gaps in the line you are tracing. When you get to a gap, click at the edge, then click at the start of the line on the other side, and continue tracing. When satisfied with the second polygon, save your edits. Finally, trace the remaining area of the third polygon as shown in the figure map at right. Keep the Editor toolbar open. You will need to be editing to do the next exercise.

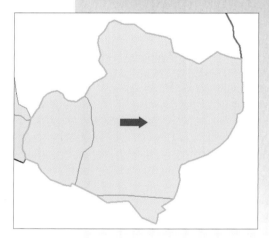

Add attribute data to new polygons

Suppose that the car beat numbers are 1, 2, and 3 in the order that you digitized them and that the car beat names are West, South, and North, respectively. In this exercise, you record these values by adding data to the attribute table of CarBeatsCustom.

1   In the table of contents, right-click CarBeatsCustom and select Open Attribute Table.

2   On the Editor toolbar, click Editor > Start Editing. Select CarBeatsCustom in the top panel. Click OK > Continue.

3   In the Create Features window, click CarBeatsCustom.

4   Click the gray row selector button for the first row and examine your map to see that the first polygon, the western one, is selected.

5   In the attribute table, click in the ID cell in the first row and type **1**.

6   Similarly, click in the ID cell in the second row and type **2**. Then click in the ID cell in the third row and type **3**.

7   In the CarBeats cell, type **West** in the first row,
     South in the second row, and **North** in the third row.

OBJECTID *	SHAPE *	ID	CarBeat	SHAPE_Length	SHAPE_Area
1	Polygon	1	West	51539.598343	105212824.909084
2	Polygon	2	South	63024.017535	190948837.031137
3	Polygon	3	North	102766.740852	241811025.90336

CarBeatsCustom

8   On the Editor toolbar, click Editor > Save Edits. Then click Editor > Stop Editing. Close the attribute
     table and save your map document.

Set up to digitize lines for block watches

Block watches are small neighborhood organizations whose members are vigilant about what goes on
in their neighborhood and who work to prevent crime. Police often provide maps of crimes for block
watch areas in support of these citizen efforts. In this exercise, you digitize a block watch consisting
of several street segments. The snapping capability in ArcMap can make your digitized lines an exact
match for the lines of streets. When you are close enough to a street segment end point or vertex with
your pointer, snapping creates a new point exactly on top of the desired point of the street segment.

1   In the table of contents, turn layers on or off so that only PennHillsStreets, PennHills, and BlockWatches
     are on.

2   Right-click PennHillsStreets and select Label Features.

3   Zoom in on the western portion of Penn Hills, and
     then zoom in closer on the small area as shown
     in the figure map.

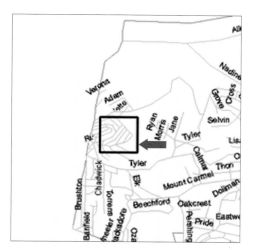

4   On the Editor toolbar, click Editor > Start Editing. Click BlockWatches in the top panel. Then click OK >
     Continue.

5   Click BlockWatches in the Create Features window.

6   Click Editor > Options.

7   Click the General tab and set "Sticky move tolerance" at **15** pixels. Click OK.

Digitize lines for block watches

Suppose the police department defines block watches by collections of street segments. In this exercise, you create block watches by snapping to existing points of streets, which are either end points of street segments or vertices that provide shape for the interior parts of street segments. As you digitize, you'll have to do some backtracking and digitize over lines that have already been created. This shouldn't pose a problem, and the resulting features will work well. The selected street segments in the figure map make up the San Juan Block Watch. You do not have to select the same streets for your block watch.

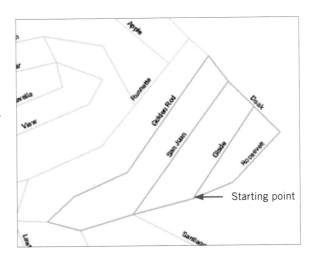

1   **Click the Trace button.**

2   **Click at the corner of Glade and Roosevelt as shown in the figure.**

3   **Move your pointer to the right along the bend in Roosevelt and click once.**

4   **Move your pointer to the end of Roosevelt and click there.** Keep adding end points and vertices until you have digitized each street of the block watch. If needed, you can click Edit > Undo on the Menu bar to remove a point that's just been digitized. You can also press ESC to escape a menu selection. If necessary, you can simply click without snapping to create a new point. When you get to the last point, place the pointer directly over the desired point and double-click to end the line feature. (You will not be able to snap to this point.)

5   **Click Editor > Save Edits. Then click Editor > Stop Editing.**

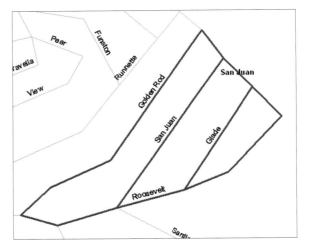

Use this figure map to create a second block watch called **North Avenue**, somewhat of your own design. Save your edits, but do not stop editing.

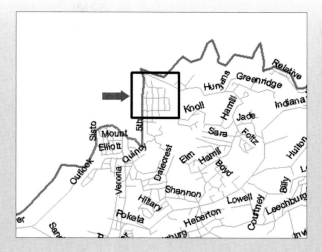

Type the numbers **1** and **2** in the ID field of the BlockWatches attribute table. Add the names **San Juan** and **North Avenue**. Label both block watches using the Name attribute in the map (use the Labels tab on the Layer Properties sheet where you can also set the placement to horizontal, parallel, curved, or perpendicular orientation). When you are finished, save your edits and save your map document. Then exit ArcMap.

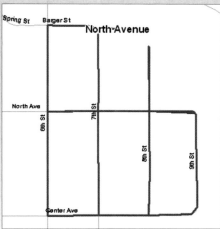

Assignment 8-1

Download and use a census block group basemap and data

For this assignment, create a new map document, add map layers that you used in this chapter, and add additional data and a map layer. Then download and process this new information. The new map layer is for census block groups. Census tracts are divided into an average of about four block groups of roughly 1,000 population each, so block groups provide a finer-grained set of polygons than census tracts for displaying census data.

Download data

Download the following data to your assignment 8-1 folder in MyAssignments:

+ Census 2000 Block Group shapefile for Allegheny County, Pennsylvania, from the Census TIGER Web site 2009 TIGER/Line Shapefiles

+ SF 3 Table P37 Sex by Educational Attainment for the Population 25+ Years for Allegheny County, Pennsylvania, block groups from the U.S. Census Bureau American FactFinder Web site

Preprocess data

Next, preprocess the SF 3 data by opening the downloaded table P37 in Microsoft Office Excel. Ultimately, the objective is to have data on the population of all males age 25 and older and the subset of males at that age with less than a high school diploma. Lack of education is one indicator of poverty, and thus of a crime-prone population. Keep only the columns in Excel as noted here and rename them using the names given in the third column:

Excel Column	Table P37 Column	New Name
A	Geo_ID2	BGID
B	P037002	MalePop
C	P037003	MaleNoHS
D	P037004	Male4
E	P037005	Male56
F	P037006	Male78
G	P037007	Male9
H	P037008	Male10
I	P037009	Male11
J	P037010	Male12NoD

If you cannot see value 420030103001 in the first cell of BGID, right-click the B column selector and click Format Cells > Number > 0 Decimal places.

The following steps provide instructions on how to create a new column in Excel that adds up data from several other columns:

1   Right-click the B column selector and click Insert.

2   In cell B1, type MaleNoHS.

3   Click in cell B2 and type
    = D2+E2+F2+G2+H2+I2+J2+K2

4   Press the TAB key.

5   Click in cell B2 and press CTRL+C to copy the cell formula.

6   Click in cell B3 and drag the pointer down through the column to cell B1108 at the bottom to select the rest of the column.

7   Press CTRL+V to paste the formula.

A sample of what you should now have is shown in the figure.

	A	B	C	D	E	F	G	H	I	J	K
1	BGID	MaleNoHS	MalePop	MaleNoSch	Male4	Male56	Male78	Male9	Male10	Male11	Male12N
2	420030103001	59	225	3			6	15	14	8	13
3	420030103002	8	65					8			
4	420030201001	121	485	11				23	31	14	42
5	420030201002	36	160			7	5				24
6	420030201003	1091	1622		15		226	151	339	315	45

Save the spreadsheet as **MaleEdAttYourName.xls** to your assignment 8-1 folder in MyAssignments.

Create new map document

Create a new map document called **Assignment8-1YourName.mxd** and save it to your assignment 8-1 folder. Make the data frame coordinate system the state plane for Southern Pennsylvania (NAD83 US Feet). Add the following map layers and data, changing names in the table of contents as indicated:

+ **Municipalities**: fe_2009_42003_cousub00.shp in the Downloads folder in chapter 8 of MyExercises
+ **Water Features**: fe_2009_42003_areawater.shp in the Downloads folder
+ **Block Groups**: fe_2009_42003_bg00.shp in your assignment 8-1 folder
+ **SF 3 Data**: MaleEdAttYourName.xls in your assignment 8-1 folder

Suppose the municipalities of West Homestead, Homestead, Munhall, and Whitaker are merging their separate police departments to form a metropolitan police department called Homestead Metro Police. Carry out the steps that follow. Save all new shapefiles to your assignment 8-1 folder.

1   Extract boundaries for West Homestead, Homestead, Munhall, and Whitaker into a single new layer, **HomesteadMetro.shp**, with state plane coordinates for southern Pennsylvania, 1983, in feet.

2 Using HomesteadMetro.shp, extract block groups for the area, **BGHometeadMetro.shp**.

3 Create a choropleth map called **Assignment8-1YourName.mxd** to display MaleNoHS normalized by Male Pop, using five quantiles and the monochromatic color ramp of your choice.

4 Clip the Monongahela River bordering on the north of the metro area, using HomesteadMetro.shp to create **MonRiver.shp**. **Hint:** Be sure to set the General Settings environment of ArcToolbox so that all output map layers are in state plane coordinates.

5 Turn HomesteadMetro.shp on and symbolize it with hollow fill and an outline width of 2. Label the municipalities by name.

6 Create a simple layout with a map, title, and legend for the census data. Save a JPEG image of your layout called **Assignment8-1YourName.jpg** to your assignment 8-1 folder.

What to turn in

Use a compression program to compress and save your assignment 8-1 folder to **Assignment8-1YourName.zip**. Turn in your compressed file.

Assignment 8-2

Create maps for foot patrols and DUI target areas

Suppose the Pittsburgh Police Bureau wants to pilot test two coordinated crime interventions: (1) foot patrols and (2) driving under the influence (DUI) target areas. Foot patrol areas partition the pilot area into smaller sections, each with a foot patrol officer assigned to specific time periods. The DUI target areas are stretches of major streets that lead to and from commercial areas that have bars and taverns.

For this assignment, create a map layer with DUI target area streets that are digitized using streets as a guide and a map for foot patrols that is created from dissolved census tracts.

Create new map document

Create a new map document called **Assignment8-2YourName.mxd** and save it to your assignment 8-2 folder in MyAssignments. Make the data frame coordinate system the state plane for Southern Pennsylvania (NAD83 US Feet). Make sure that all layers you create have this coordinate system.

Add the following map layers and data:

+ **Streets:** Streets in the Pittsburgh geodatabase in the Data folder
+ **Census Tracts:** Tracts in the Pittsburgh geodatabase
+ **Commercial Areas:** ZoningCommercialBuffer in the Pittsburgh geodatabase
+ **Rivers:** Rivers in the Pittsburgh geodatabase

As soon as you are finished with any of these input map layers and it is no longer needed, remove it from your map document.

Extract pilot study map layers

Create and save a new file geodatabase called **Assignment8-2.gdb** to your assignment 8-2 folder and save all new map layers to it.

The figure map shows selected study area census tracts. Using these census tracts, create the following map layers:

+ **Pilot Census Tracts:** Tracts in the Pittsburgh geodatabase

+ **Pilot Streets:** Streets in the Pittsburgh geodatabase

+ **Pilot Rivers:** Rivers in the Pittsburgh geodatabase

+ **Pilot Commercial Areas:** ZoningCommercialBuffer in the Pittsburgh geodatabase

Hint: Clip the commercial areas using the pilot census tracts.

Symbolize the finished map layers. **Hint:** For ZoningCommercialBuffer.shp, use the following steps:

1   Open the property sheet for this layer and click the Symbology tab.

2   Click Categories > Unique Values > Add All Values.

3   Symbolize using 50% transparency and see if you can use a line fill symbol. (Click Edit Symbol in the Symbol Selector and select Type.)

Dissolve census tracts for foot patrol areas

The figure map included here shows foot patrol areas in terms of census tracts. The four proposed foot patrol areas are each composed of multiple census tracts. Dissolve the census tracts to produce foot patrol polygons called **FootPatrols** saved to the assignment 8-2 geodatabase.

Digitize DUI target areas

Major streets have CFCC values that are less than or equal to the value in cell A35 in the Streets attribute table (see the figure map, which has these streets selected). Create two DUI target areas (polygons) by identifying two to four blocks of major streets that are within zoning commercial buffers. Digitize a single polygon for each DUI area with lines made up from the blocks on either side of identified street segments (so that the DUI target areas are two blocks wide with identified street segments in the center). Name the individual areas **DUI 1** and **DUI 2** and use these values to label the DUI target areas on the map. Save the results as **DUITargets** to the assignment 8-2 geodatabase. **Hint:** Use the Trace tool.

What to turn in

Use a compression program to compress and save your assignment 8-2 folder to **Assignment8-2YourName.zip**. Turn in your compressed file.

Chapter 9

Preparing incident data for mapping

The constant flow of new incident data that needs to be processed into spatially enabled formats and then mapped is a distinguishing feature of crime mapping. The needed GIS work uses the highly sophisticated process of geocoding, which transforms incident report street addresses into points on a map. Additional GIS work, spatial overlay, assigns polygon identifiers to geocoded points and ultimately yields aggregate data, such as monthly crime counts by police sector. You can think of the work in this chapter as building a spatial data warehouse in which all the current and past crime incident data comes from computer-aided dispatch, offense reports, and arrest reports, as well as other specialized data. All this geocoded data is then ready for crime mapping and analysis.

Data processing and geocoding concepts

Police reports generally include location data—street addresses such as 3215 Liberty Avenue, intersections such as Forbes Avenue and Craig Street, or place names such as Pittsburgh International Airport. The GIS task is to transform this location data into the corresponding x- and y-coordinates on Earth's surface through the use of basemaps and sophisticated GIS computer algorithms. This process is known as geocoding, or address matching. In this introduction, you learn about geocoding, and then later in the chapter, you try out the tools in ArcMap to geocode addresses yourself.

Master and update files

An important function of the crime mapping and analysis unit of a police department is to process the daily flow of new crime incidents into master files. Typically, the database system used for an organization's operational data keeps master files of incidents that span several years and are updated with the new data that comes in. New data, perhaps from the previous day or an earlier shift, is placed in an update file, processed into the desired format, and then appended to the master file. The master file that contains all past and present incident data is available for analytical reporting, ad hoc queries, mapping, forecasting, and detailed crime analysis—meeting the information needs of the organization.

Police reports and location data

The three primary sources of police incident data are emergency calls for service taken by computer-aided dispatch (CAD), and offense and arrest reports taken by police officers and detectives. Secondary sources may include field contact reports, surveillance reports, and property recovery records. All of this incident data is collected on a systematic basis with the use of various forms. Each of the corresponding records may have one or more addresses, which are generally entered in a free-form text field without the benefit of automated data validation rules that check for accuracy or consistency.

The resulting address data is often far from perfect, with different people using different spellings for the same street names; different abbreviations for the same street types, such as Av, Ave, or Avenue; and different ways of noting addresses, such as by block rather than by specific address as in the 500 block of Main Street. In addition, sometimes an officer may include extraneous notes in the address data entry fields that are not part of the actual address, such as "vicious dog in backyard." Geocoding has the capacity to produce useful crime maps, even though the source data is imperfect. Before seeing how geocoding in ArcMap handles imperfections in address data, you need to take a look at the available reference data, TIGER street centerlines, used to assign coordinates to street addresses.

TIGER street centerlines

The primary spatial-reference data used for geocoding in the United States is the TIGER/Line data developed by the U.S. Geological Survey and the U.S. Census Bureau. The figure illustrates TIGER (Topologically Integrated Geographic Encoding and Referencing) street centerlines for a selected street segment in Pittsburgh. You can see that the segment is a block long and that it has only the starting and ending street numbers for the odd side of the street (3201 and 3299) and the even side of the street (3200 and 3298). The crime point (red point marker) is for an incident at 3215 Liberty Avenue and is offset 20 ft on the odd side of the street (northwest, in this case). The point was located by the use of geocoding, so it was interpolated to be 15% of the distance between the start and end of the block. The actual location of that address is likely different from the interpolated point that is plotted.

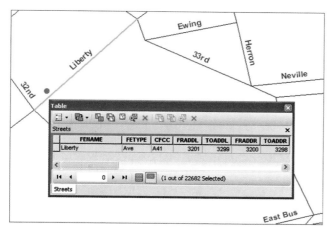

Example of a TIGER street centerline segment and record.

Geocoding of address data

The geocoding algorithms in ArcMap are designed to handle imperfect source data, and thus use a "fuzzy" matching process in which source and reference data do not have to match perfectly to be considered a match. These algorithms use a rule-based expert system that has a scoring mechanism to rate potential matches of reference addresses to a source address. The reference address with the highest score above a minimum acceptable score is deemed the match for a source address, even if it is not 100% accurate.

Sometimes, the address field of a report has a place name (for example, PNC Park) instead of the location address (115 Federal Street). ArcMap has the capacity to use an alias file that includes all the place names for a selected address. For example, through aliases, PNC Park, PNC Ballpark, and PNC Pirates Park can all be linked to 115 Federal Street. ArcMap makes a pass through source addresses, searching for aliases and replacing them with corresponding street addresses before address matching. In some cases where source data has specific patterns that cannot be handled by ArcMap or with an alias table, you can perform "data cleaning" in Microsoft Office Excel software or custom programming before address matching. For instance, you could replace "500 block of Main St" with an approximation such as "550 Main St," assuming that blocks are 100 units long.

Address matching errors and reporting

Errors are bound to occur when using the fuzzy matching in address matching as a trade-off for not matching records at all. This brings us to the limitations of crime mapping. Mapped crime incidents that have some degree of error are still useful for identifying and diagnosing spatial and temporal patterns of criminal activity. For example, even if some crime points were in error, the presence of 25 mapped incidents in the same block over a month would generally indicate a problem that called for detailed crime analysis, including the information gleaned from reading police reports and talking to field officers and informants. Crime maps also serve to integrate a variety of data for field officers. In another example of maps that are useful even if they have errors, the maps can be used by police officers to get information on crimes that occurred in their patrol sectors on shifts other than their own, on days when they were off, or in areas that span adjoining beats.

On the other hand, a point that is address matched would not be considered accurate enough to serve as legal evidence of a crime scene. Because ArcMap interpolates location through the use of TIGER street numbers, a point plotted on a TIGER map may not be on the correct land parcel, given TIGER limitations. Moreover, TIGER street centerline maps have errors in the locations and shapes of line features. Hence, the general role of crime maps is to portray crime patterns, despite the tendency of these maps to have errors of many kinds.

Even so, there are ways to improve address matching rates, and address match errors can be resolved. A policy that all incident addresses either (1) have a standard street address or (2) have a street intersection (for example, Forbes Ave & Craig St), or (3) are a place name that appears on the official list being used will yield good matching results given a good reference street map—that is, have an address matching rate in excess of 95%. One remedy for nonmatches or errors is to provide verification feedback loops on the process of recording addresses. For example, if data entry is decentralized and done at the precinct level, the local crime analyst should audit each address (and other data elements) as they are entered and communicate any problems to the officers writing reports so corrections can be made. (This is the current practice of the Pittsburgh Police Bureau.) Also, any gaps or errors in the reference map itself can be corrected in ArcMap by editing attributes or features.

Crime data aggregation

Once an update file is address-matched as much as possible, the results can be appended to the master incident file. Then when the master file is up-to-date, aggregate data can be used for crime analysis. For example, monthly crime counts by car beat or census tract are derived from aggregate data. Space and time series aggregate data can be used to detect crime flare-ups, estimate the seasonal effects of crime, and assess long-term crime trends.

Geocoding crime incident data

In the following exercises, you'll carry out batch and interactive address matching, append the results to a master file, carry out spatial overlay of selected data to assign car beat IDs to incident points, aggregate data to space and time series data, and protect the privacy of data for public use by "fuzzing" location addresses. You'll also carry out many additional smaller tasks to improve the rate of address matching.

Tutorial 9-1

Address matching, or geocoding, data

The task in this set of exercises is to match the address data in source police data to the addresses in reference street maps. ArcMap associates the source address data with a reference street segment and places a point on the street that becomes the approximate, if not necessarily exact, location of the incident. And thus, a street address is geocoded as a point on the map.

Open map document

1   **Open Tutorial9-1.mxd in ArcMap.** This map of Pittsburgh has only the map layers that are essential for geocoding—namely, Streets and PoliceSectors, and the addition of Pittsburgh's municipal boundary, which was included to make the map look complete. You need Streets as reference data for address matching and PoliceSectors as spatial overlay for adding area IDs to the matched points. The map document also has the source table for addresses, OffenseMaster20081031$.

2   **Save your map document to your chapter 9 folder in MyExercises.**

Create new address locator

ArcGIS saves settings or parameter values used for carrying out address matching to reusable files called locators. With the use of Catalog, you can create a locator for Pittsburgh TIGER streets.

1   **On the Standard toolbar, click the Catalog window button.**

2   **In the Catalog window, go to your chapter 9 folder in MyExercises.**

3   **Right-click the chapter 9 geodatabase and click New > Address Locator.**

4 In the Create Address Locator dialog box, click the Browse button for Address Locator Style, and then scroll down and click US Address - Dual Ranges. Click OK and ignore the warning icon and message in the Create Address Locator dialog box. US Address - Dual Ranges is the style used for TIGER street centerline maps, which have ranges of street numbers (starting and ending) for the left and the right sides of the street.

5 **Click the Reference Data arrow and select Streets.** The Field Map of the Create Address Locator dialog box automatically identifies all the fields it needs in the street map for the geocoding process. If ArcMap were unsuccessful in identifying needed fields, you would have to click in a cell on the right of an address component and select the needed field name from the table. Field names that start with an asterisk are required for geocoding addresses. **Note:** If you have TIGER streets for a task force or jurisdiction that has more than one city, you need a reference map that includes "left city" and "right city" attributes for each street segment so that you can correctly match streets that have commonly used names. For example, 123 Main St., Pittsburgh, is not distinguishable from 123 Main St., Penn Hills, without the city name. If you do not have city in your reference map and in your source data, ArcMap arbitrarily uses the first match that it finds for 123 Main St., which leads to approximately 50% address matching errors for this scenario. In the current data, there is only one city in the police jurisdiction, Pittsburgh, so you do not need to designate it in the locator field map as left city or place and right city or place.

6 Scroll down in the Create Address Locator dialog box, click the Browse button for Output Address Locator, go to your chapter 9 folder in MyExercises, and double-click the chapter 9 geodatabase. Type **PittsburghTIGER** for Name and click Save.

7 Click OK and wait until Catalog informs you that the address locator is created. Catalog creates the indices needed for the address matching process.

8 Save your map document.

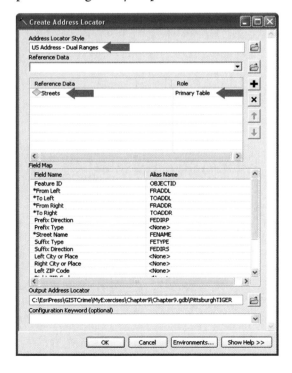

Modify address locator properties

1 **In the Catalog tree, expand the chapter 9 geodatabase and double-click PittsburghTIGER.** Now you have the Address Locator Properties sheet. First, notice that you can associate a place name alias table with the address locator (see the button in the Matching Options panel). Next, notice that the allowable connectors for street intersection addresses, such as Oak St & Pine Av, are currently the characters &, @, and | and the word *and* . If your source data has a different connector, you could type it here. Also, the point that ArcMap assigns to an address has a 20 ft offset on the correct side of the street (left or right). You can change the offset to another value if desired. Administrative boundaries such as police sectors are mostly made up of streets, and a side offset guarantees that address points fall into the correct polygons when crimes are counted by area. Otherwise, points that have no offset would be double-counted, once for each polygon sharing a street boundary, which would lead to address matching errors.

2 **In the Address Locator Properties dialog box, in the Matching Options panel, click Place Name Alias Table.** You will add more aliases to an alias table in a later exercise, but for now, you can use what's supplied in the book's data files. The current alias table is provided in the figure for your information.

3 **In the Alias Table dialog box, go to the Locators folder in Data and double-click PittsburghAlias.csv.**

4 **In the Alias field, select Alias. Click OK.**

Alias	Address	City
ALLDERDICE HS	2409 Shady Ave	Pittsburgh
FRICK PARK	1901 Beechwood Blvd	Pittsburgh
GATEWAY PLAZA	1600 W Carson St	Pittsburgh
GREYHOUND BUS STATION	1151 Penn Ave	Pittsburgh
HEMLOCK GARAGE	200 Hemlock ST	Pittsburgh
HEMLOCK STREET GARAGE	200 Hemlock ST	Pittsburgh
HOMEWOOD PARK	557 N Lang Ave	Pittsburgh
MON WARF	101 Fort Pitt BLVD	Pittsburgh
MON WARH PARKING GARAGE	101 Fort Pitt BLVD	Pittsburgh
MON WHARF	101 Fort Pitt BLVD	Pittsburgh
POINT STATE PARK	251 Commonwealth PL	Pittsburgh

5 **Enter selections in the Address Locator Properties dialog box as shown in the figure.** Many are automatically selected for you. In the Matching Options panel, notice the three parameters under the Place Name Alias Table button. These are the default threshold values (80, 10, 85) that control how closely the spelling of the street address in the police reports table must be to match that of the TIGER street map. It is not possible to interpret these values with any precision. However, if they are too low (lenient), the result will be numerous incorrect matches. If they are too stringent, too many police report incidents will go unmatched, and thus not be mapped. Here, simply

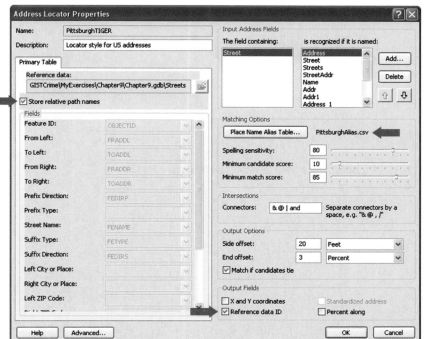

use the defaults. Note that trial and error is the best approach to selecting the best parameters. One approach to selecting parameters is to build alternative address locators that use strict (high) to lenient (low) values for matching options. Then examine the additional records that are matched when more lenient locators are used. Visual inspection of the attribute table for geocoded addresses using the more lenient locator will reveal the point where the source address of records no longer matches the address that ArcMap assigns, serving as an indicator of where to draw the line. When the address locator is too lenient, the "matched" addresses are clearly wrong.

6   Click OK to close the Address Locator Properties dialog box and click Auto Hide to close the Catalog window. Save your map document.

Batch address match incident data

You are now going to work with a large volume of data: all 51,626 offense records for the City of Pittsburgh from January 1 through October 31, 2008. It would take your computer approximately 20 minutes to do this massive amount of work. Instead of having to wait for processing to complete, you will start geocoding to go through the steps, and then cancel the operation. Then you will add a copy of the geocoded result from the chapter 9 geodatabase to your map document so that you can carry out additional work in the chapter. If you wish to let geocoding finish on your computer, you can skip the steps in this exercise for adding the provided geocoded map layer. Later in this chapter, you will geocode a smaller amount of source data to completion.

1   From the Menu bar, click Customize > Toolbars > Geocoding. The free-floating Geocoding toolbar appears on your computer desktop.

2   On the Geocoding toolbar, click the Geocode Addresses button 🔖 and select PittsburghTIGER. Click OK. ArcMap automatically fills in OffenseMaster20081031$ as the address table in the Geocode Addresses: PittsburghTIGER dialog box, because it was included in the map document, Tutorial9-1.mxd.

3   In the "Output shapefile or feature class" field, go to the chapter 9 geodatabase, type **OffenseMaster20081031** for Name, and click Save.

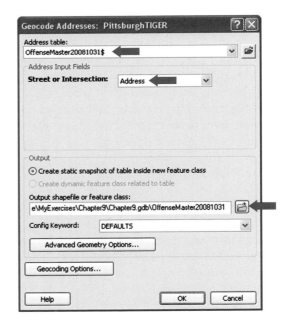

4   Click OK to batch address match all input police reports.

5   After processing for a minute or two, click Cancel > Close and remove the partially completed Geocoding Result layer from the table of contents.

6   Add OffenseMaster20081031b from the chapter 9 file geodatabase in MyExercises.

7   Right-click OffenseMaster20081031b in the table of contents, click Data > Review/Rematch Addresses to open the Interactive Rematch dialog box. You will use this interface in the next exercise to match some of the unmatched addresses. For now, you can see the current match statistics: 42,373 (82%) matched (a lot of points), another 684 (1%) tied for best score but also with a match chosen, and 8,569 (17%) unmatched. So, 83% of the offenses are matched and on the map. Some of these offenses will, of course, have location errors, even though they are plotted on the map.

8   Close the Interactive Rematch dialog box.

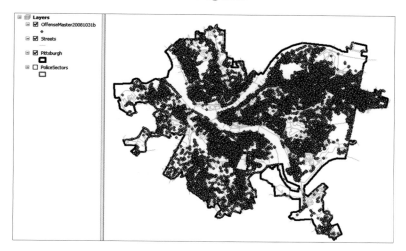

Tutorial 9-2

Improving address matching results

Although many things can go wrong in address matching, a lot can be done to improve the results. Knowing the many approaches to improve address matching can give you a head start on becoming a GIS specialist. Crime mapping and analysis depends on comprehensive and accurate mapping of police reports—your current task in this exercise.

Refine address matching statistics

In this exercise, you examine the address matched layer's attribute table and diagnose the unmatched offenses. Some address fields have missing or blank values and thus are impossible to match. Others do not have matchable address data entries, which must start with a number (for example, 123 Oak St), be a street intersection (for example, Oak St & Pine Av), or be a place name in the alias table.

1   Save your map document as **Tutorial9-2.mxd** to your chapter 9 folder in MyExercises.

2   In the table of contents, right-click OffenseMaster20081031b and select Open Attribute Table. Scroll to the right, right-click the column heading Address, and click Sort Ascending. You can now select or filter records to view subsets of all records. Note that the records on display have no addresses entered and so are impossible to geocode. In this case, each of these records has a blank space entered in the Address field.

3   From the Menu bar, click Selection > Select By Attributes.

4   In the Select By Attributes dialog box, scroll down in the panel with field names, double-click "Address", click the = button, and type a space, a single quotation mark, a space, and another single quotation mark. The result is an SQL criterion for filtering records ("Address" = ' '), which selects all records that have a blank value for Address.

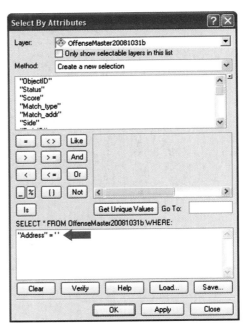

5   Click OK. Save your map document, but do not close it. At the bottom of the attribute table, you can see that 1,805 (3.5%) of 51,626 records are selected, and thus have missing addresses. Clearly, it is not the fault of ArcMap or the Streets layer that these records are not geocoded. Rather, it is because their address fields were blank.

Count records with unmatchable address formats

Some records are not matchable because their street address data does not have a street address or intersection format. Address field values that do not start with a digit are not street addresses, because by definition, an address starts with a house number. Similarly, address field values that do not contain an ampersand (&) are not intersections, because the Pittsburgh data uses the ampersand as the connector between two street names for an intersection. That leaves only one way that these records can be matched—that is, if they are a place name in an alias table. So now, you can execute a sophisticated SQL query to find nonmissing records that were not geocoded because they are not an address or an intersection, but could be a place name. You build the following query in steps 1–3: "Status" = 'U' AND "Address" NOT LIKE '%&%' AND "Address" > '99999'. The explanation follows.

1   From the Menu bar, click Selection > Select By Attributes.

2   Click the Clear button.

3   Double-click "Status", click the = button, and type a space and **'U'** (including the single quotation
marks). Next, click the And button. Double-click "Address", click the Not button, click the Like button,
and type a space and **'%&%'**. (Be sure to type the single quotation marks.) Next, click the And button.

Finally, double-click "Address", click the >
button, and type a space and **'99999'**. The
query selects all records that are unmatched
("Status" = 'U'), have an address that does
not contain an ampersand ("Address" NOT
LIKE '%&%'), and do not start with a number
("Address" > '99999'). LIKE is an operator
that looks for substrings or parts of text
values. Here, % is a wild card representing
0, 1, or more characters followed by an
ampersand and followed by 0, 1, or more
characters as represented by the second
wild card. So this part of the query looks
for values that have an imbedded &. **Note:**
"Address" > '99999' is a "trick" of geocoding.
When text values are sorted in ascending
order, values starting with numbers are listed
first. Here, it is assumed that no address has a
street number greater than 99999. The result
of this part of the query is address values that
do not start with a number.

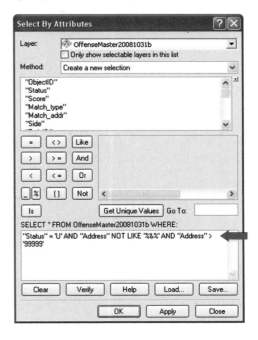

4   **Click the Verify button.** If you do not have a syntax error, you get a message that says your
expression was successfully verified. If you have an error, study the finished expression that
appears in the text before step 1, check your query expression, and correct any errors. Try
verifying your query until you get it right.

5   **Click OK twice. Then in the attribute table, click the "Show selected records" button** ⊟ **at the bottom
of the table to show only the selected records.** There are 545 (1.1%) values selected out of the total
51,626. Some of these records have place names instead of addresses (for example, ALDERDICE
HIGH SCHOOL), which you can place in the alias table and then fill in the street addresses.
While helpful, this is not going to lead to a large improvement in match statistics for Pittsburgh's
offense data (although it might in other jurisdictions). If you take the 545 + 1,805 = 2,350
impossible records out of the total 51,626, the match rate of the remaining source record
addresses and reference data TIGER map is 100*(42,532 + 684)/(51,626 − 2,350) = 88.0%. This
match rate is typical of police offense data in the United States.

6   **From the Menu bar, click Selection > Clear Selected Features. Close the attribute table.**

Clean data with spreadsheet software (optional exercise)

If there are address entries that cannot be geocoded and have recognizable problems, you can modify or "clean" the data automatically by using Excel or your own custom program so the data can be geocoded. For example, many police departments use some "block of" address data, such as the 500 block of Main Street. While this is not a practice in Pittsburgh, suppose that the following records were selected with a query:

300 BLOCK OF FREELAND ST	400 BLOCK OF LIBERTY AV
7000 BLOCK OF BENNETT ST W	100 BLOCK OF GRANT ST

In this exercise, you remove "BLOCK OF" and then add 50 to street numbers, assuming that the blocks are 100 numbers long, to yield approximate street addresses in the middle of a block.

1   Start Excel. Type the preceding block of addresses into cells A1, A2, A3, and A4. Click the Home tab and, in the Editing panel, click Find & Select > Replace.

2   In the "Find what" text box, type **BLOCK OF** and press the SPACEBAR. Click Replace All > OK > Close.

3   Select cells A1 through A4, click the Data tab, and in the Data Tools panel, click the Text to Columns button.

4   In step 1 of the Convert Text to Columns Wizard, click the Delimited option and click Next. In step 2, click to clear the Tab check box and select the Space check box. Click Next. In step 3, click Finish. Excel parses the A column text values into four columns, using a space as the delimiter (or separator for columns).

◢	A	B	C	D
1	300	FREELAND	ST	
2	7000	BENNETT	ST	W
3	400	LIBERTY	AV	
4	100	GRANT	ST	

5   Right-click the A column heading and click Insert. Then repeat this step to create two new blank columns, A and B.

6   Click in cell B1 and type **=C1+50**. Press ENTER. Click in cell B1, and then press and drag the handle at the lower-right corner of the cell down through cell B4.

7   Right-click the B column heading and click Copy. Then right-click the B column heading, click Paste Special, and click the Values option. Click OK.

8   Right-click the C column heading and click Delete. Then click in cell A1 and type **=B1 & " " & C1& " " & D1 & " " & E1**. Press ENTER. Click in cell A1, and then press and drag the handle in the lower-right corner of the cell down through cell A4.

9   Right-click the A column heading and click Copy. Then right-click the A column heading, click Paste Special, and click the Values option. Click OK.

10   Delete columns B through E, take a look at the results (the addresses are now address-matchable), and close Excel without saving.

◢	A
1	350 FREELAND ST
2	7050 BENNETT ST W
3	450 LIBERTY AV
4	150 GRANT ST

Add more alias values

The last query you ran in ArcMap does a good job of identifying potential place names for an alias table. All you have to do is scroll through the selected records and look for addresses that have place names. (The place names would be known to you if you lived in Pittsburgh.) Then you can add the place names and their addresses to the alias table and rebuild the PittsburghTIGER locator. One way to get addresses for place names is to use the Internet to look them up; for example, try superpages.com, MapQuest.com, or maps.google.com and search for a place name, using "Pittsburgh, PA" for the city and state. Some potential new place names from the selected records include:

ALDERDICE HIGH SCHOOL	EAST HILLS PARKING LOT
ALLEGHENY CENTER MALL	FRICK PARK NATURE CENTER
ALLIES PARKING GARAGE	HEINZ FIELD ART ROONEY DR
ANDERSON PLAYGROUND	MON WHARF PARKING LOT
CCAC ALLEGHENY CAMPUS	MONTIFIORE HOSPITAL
CHATHAM COLLEGE	PARKING LOT PARKWAY CENTER DR
CHILDRENS HOSPITAL PGH	PEABODY HIGH SCHOOL
CLARK BLDG	PNC PARK
CLEMENTE PARK	STATION SQUARE
COMMONS STREET BUILDING	VIETNAM VETS
E HILLS MALL PARKING LOT	WEST END OVERLOOK PARK

Names of bridges and other locations that do not have street numbers or intersections cannot be handled directly through aliases. You have to modify the TIGER map to handle these cases, which you do will later in this chapter. For now, you can add a few straightforward aliases to the alias table—namely, ALDERDICE HIGH SCHOOL, CLARK BLDG, MONTIFIORE HOSPITAL, and PNC PARK. "MONTIFIORE " is misspelled and should be "MONTEFIORE," so you can add two lines for the place name, one spelled correctly and the other misspelled. Also, ALDERDICE is misspelled and should be ALLDERDICE, but an alias with the correct spelling is already in the alias table, so you can add just the misspelled version. You can do this work with the Microsoft Notepad text editor. If you were to count the number of records in the geocoding results that have these selected place names, you would find 11. So, you can expect to add 11 newly matched records as a result of the expanded alias table once you rematch the records. Now you will edit that table.

1   From the Windows Start menu, click All Programs > Accessories > Notepad.

2 In Notepad, click File > Open. Change "Files of type" to All Files, go to the Locators folder in Data, and click PittsburghAlias.csv. Click Open. The data format is comma-separated values (.csv). Instead of the data values being in separate columns, the values are separated by commas. You can see that there is already an entry for Allderdice High School in the table—and that it has the correct spelling. So, you'll only need to add one row for the second, incorrect spelling of Alderdice High School, but use the same address, 2409 Shady Ave.

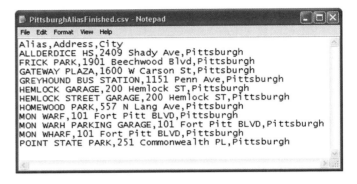

3 In Notepad, click at the end of the line with ALLDERDICE HS, press ENTER to open a new line, and type **ALDERDICE HIGH SCHOOL,2409 Shady Ave,Pittsburgh.** (Alternately, first select the existing ALLDERDICE HS line and press CTRL+C to copy the line. Then click at the end of the same line, press ENTER, press CTRL+V to paste, and change ALLDERDICE to **ALDERDICE** and HS to **HIGH SCHOOL**).

Although you could use a variety of Web sites to find the street addresses for place names, in the next step, use the following results:

CLARK BLDG,717 Liberty Ave,Pittsburgh

MONTEFIORE HOSPITAL,3459 5th Ave,Pittsburgh

MONTIFIORE HOSPITAL,3459 5th Ave,Pittsburgh

PNC PARK,115 Federal St,Pittsburgh

4 Enter the preceding values into Notepad, roughly alphabetized with the existing entries.

5 Save your text file to your chapter 9 folder in MyExercises. Close Notepad.

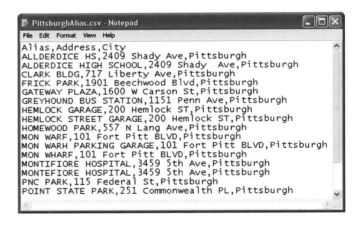

Verify addresses in the alias table

Before using the alias table, you should use the Find tool in ArcMap to see if the addresses you have found for aliases can be geocoded and whether the resulting points on the map are correct.

1 If you do not see the Tools toolbar, click Customize > Toolbars > Tools.

2 On the Tools toolbar, click the Find button.

3 Move the Find window if necessary so you can see the map.

4 In the Find window, click the Locations tab. Go to the chapter 9 geodatabase in MyExercises and add the PittsburghTIGER locator. Then type the Clark building address, **717 Liberty Ave**, for Full Address and click Find. The Find tool briefly places a point marker on the map at the cited address. If you can find an address with the Find tool, you know your address locator will be able to find it in address matching. Note the score of 100, indicating a perfect match.

YOUR TURN **YOUR TURN** YOUR TURN YOUR TURN YOUR

Change the name of BurglaryJuly2008Graduated to **Burglary Count: July 2008**. Also for the same layer in the table of contents, click ICOUNT, wait a few seconds, click it again, and delete the "I," resulting in the name COUNT.

Edit streets to improve address matching

You can wait on using the extended alias table to rematch addresses until you have fixed some of the issues with the reference data—that is, the street map. Then you will need to rebuild the PittsburghTIGER locator so that it can find your street file repairs. The task in this exercise is to add street numbers to locations such as bridges or interchanges, which generally do not have street numbers, or to other features in the Streets layer that are simply missing street numbers. You can use a convention for "synthetic" street numbers between 1 and 100 for bridges, which usually do not have street numbers, and then use any number in that range to allow address matching. You could also make entries in the alias table—for example, LIBERTY BRDG, 50 Liberty Brg. **Note:** the police data uses "BRDG" to designate a bridge, but "Brg" is more prevalent in ArcMap.

1 In the table of contents, turn the geocoded point layer off and at the top, click List By Selection. Click the toggle key to the right of each layer except Streets, so that Streets is the only selectable layer.

2 At the top of the table of contents, click List By Drawing Order.

3 On the Tools toolbar, click Zoom In and zoom in on the area shown in the figure map.

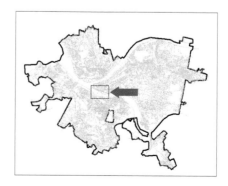

4 In the table of contents, right-click Streets and select Label Features.

5   On the Tools toolbar, click Select Features and click in the center of Liberty Bridge to select a segment of the bridge.

6   In the table of contents, right-click Streets and select Open Attribute Table. Click "Show selected records" to see the selected bridge segment.

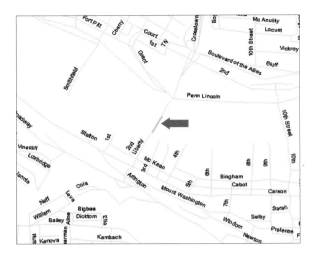

7   From the Menu bar, click Customize > Toolbars > Editor.

8   On the Editor toolbar, click Editor > Start Editing. Select Streets. Click OK. Then close the Create Features table.

9   In the selected attributes of the Streets table, type the following numbers for indicated attributes:

FRADDL	1
TOADDL	99
FRADDR	2
TOADDR	100

Adding these numbers adds the "synthetic" street numbers 1–99 odd numbers on the left side of the bridge line segment and 2–100 even numbers on the right side. To match the address 50 Liberty Brg in this segment, ArcMap will interpolate halfway along the right side of the segment to locate the point.

10   Click Editor > Save Edits.

Although you should edit the Smithfield Bridge, to the left of the Liberty Bridge, to extend it to Carson Street so it does not stop in the middle of the river, leave it as is. Instead, give the existing segment "synthetic" street numbers 1–99 odd numbers and 2–100 even numbers. When you are finished editing, save the edits and close the Streets attribute table and the Editor toolbar.

Then add the following lines to the PittsburghAlias.csv file:

LIBERTY BRDG,50 Liberty Brg,Pittsburgh

SMITHFIELD ST BRDG,50 Smithfield Brg,Pittsburgh

There are three records with Liberty Brg as an address and one with Smithfield Brg, so you can expect to add four more matched addresses, thanks to the new bridge aliases.

Although it is possible to rebuild your existing address locator, you can simply build a new one. Regardless of whether you rebuild the existing locator and rematch the current geocoding results or build a new address locator and rematch from scratch, ArcMap processes all the source data and takes the same amount of time. So now, build a new locator called **PittsburghTIGER2** and save it to the chapter 9 geodatabase. Again, use US Address - Dual Ranges for Address Locator Style and Streets for Reference Data. Open PittsburghTIGER2 in Catalog and use the PittsburghAlias.csv table in your chapter 9 folder in MyExercises—the table you just modified.

Rematch addresses

With the improvements made in the previous exercise, you should be able to get more addresses to match in OffenseMaster20081031. If you do not have 20 minutes or more to wait for processing to complete, you can skip the steps in this exercise.

1 In the table of contents, right-click OffenseMaster20081031b and click Remove.

2 On the Geocoding toolbar, click the Geocode Addresses button. Make sure PittsburghTIGER2 is selected and click OK.

3 In the "Output shapefile or feature class" field, click Browse and type **OffenseMaster20081031d** for Name. Click Save > OK.

4 After a minute or two, click Cancel > Close and remove the partially completed Geocoding Result layer from the table of contents. The corresponding, fully geocoded results are available on file, which you add next.

5 Add OffenseMaster20081031c from the chapter 9 file geodatabase in MyExercises.

6 Right-click OffenseMaster20081031c in the table of contents, click Data, Review/Rematch Addresses. Now you can see that 42,389 + 684 records are matched, 16 more as expected from the changes you made to the alias table and the streets layer.

7 Leave the Interactive Rematch dialog box open.

Rematch addresses interactively

Using general knowledge of local streets and some additional research, it is possible to match some of the unmatched addresses by using the Interactive Rematch interface. In the following steps, the "local expertise" is provided, and all you need to do is use the interface to make additional matches.

1   At the top of the Interactive Rematch dialog box, click the "Show results" arrow and select Unmatched Addresses.

2   Scroll to the right in the Interactive Rematch dialog box, right-click the Address column heading, and click Sort Ascending.

3   Scroll down and click in the first of two rows with 1 FARHAM Ct in the Address column. The street name is misspelled, missing the "e" in Fareham.

4   In the "Street or Intersection" field in the lower left panel of the Interactive Rematch dialog box, type **E** to get 1 FAREHAM CT and press the TAB key. The Interactive Rematch dialog box finds all the street address candidates and highlights the best, a 100 score match.

5   Click the Match button at the bottom of the window.

6   Click the next record and repeat steps 4 and 5. The matched total is now raised by two, to 42,391. Note that although interactive matching is impractical for the more than 50,000 records you are processing, it is a practical procedure for use on a daily basis in a police department. For a city of Pittsburgh's size, when address matching an update table of only a day's worth of offense reports (on the order of 100 to 200 records), you can expect, at most, 20 to 40 unmatched records. That's a small enough number to try interactive rematching and attempt a match rate of 100%

Rematch an address by clicking the map

The Interactive Rematch interface makes it possible to click a point on the map to geocode it, which can be an excellent option for getting results.

1   Scroll down in the Interactive Rematch dialog box and click the first record for 2 HOT METAL ST.

2   Minimize but do not close the Interactive Rematch dialog box. Hot Metal Street is not listed in the Streets map layer, but it is an alternate name for 29th Street in the block that intersects Carson Street. You can find the location on the map, and then use the Interactive Rematch interface to click a point to geocode the record.

3   On the Tools toolbar, click Full Extent.

4   Click Find, and then click the Locations tab. Under "Choose a locator," go to the chapter 9 geodatabase in MyExercises, select PittsburghTIGER2, and click Add.

5   Type **Carson ST & 29th ST** in the Full Address field. Move the Find window away from the map if necessary and click Find to see the location. You can click the Find button as many times as needed so that you can zoom in, in the next step.

6   In the panel at the bottom of the Find dialog box, right-click the matched intersection, E Carson St & S 29th St, and click Zoom To.

7   Close the Find dialog box.

8   Restore the Interactive Rematch dialog box and make sure the record for 2 HOT METAL ST is still selected. Move the Interactive Rematch dialog box out of the way of the map, and click the Pick Address from Map button.

9   Position the resulting pointer over the map at the intersection of Carson and 29th streets, right-click the intersection, and select Pick Address. If necessary, turn on Geocoding Result: OffenseMaster20081031 to see the point. The offense location is now on the map. Scroll to the left for the record of 2 HOT METAL ST and note that it is matched by use of Match_type PP (for picked point).

10   Repeat steps 8 and 9 for the second record for 2 HOT METAL ST. When finished, close the Interactive Rematch dialog box.

Tutorial 9-3

Processing update and master data files

ArcMap has all the functionality for handling the flow of new data into GIS. The powerful ArcToolbox tools and geoprocessing algorithms in ArcMap can help you quickly update daily crime reports. The following exercises go through the steps to update a master point feature class to include all the crimes for a given day.

Batch address match update offense file

Assume that it's the end of the day, November 1, 2008, and new offense data, OffenseUpdate20081101.xlsx, is ready for address matching and appending to the master feature class, OffenseMaster20081031c. The result is OffenseMaster20081101.

1   Save your map document as **Tutorial9-3.mxd** to your chapter 9 folder in MyExercises.

2   On the Tools toolbar, click Full Extent and turn the Geocoding Result Map layer off. Now, it is time to make a working copy of the current master file, OffenseMaster20081031c, called OffenseMaster20081031Temp. That way, if anything goes wrong in appending, you still have the original master file to try again. It is best practice to use a working copy when working with master files.

3   Click Auto Hide to get Catalog and go to the chapter 9 geodatabase in MyExercises.

4   Right-click OffenseMaster20081031 and click Copy. Right-click the chapter 9 geodatabase and click Paste. Click OK. Then rename OffenseMaster2081031_1 to **OffenseMaster20081031Temp**.

5   Add OffenseMaster20081031Temp to your map document (you can drag it from Catalog to the table of contents).

N YOUR TURN   YOUR TURN   YOUR TURN   YOUR TURN   YOU

Add OffenseUpdate20081101$ to your map document (go to the OffenseUpdate20081101 spreadsheet in the RawData folder) and batch address match the file by using your PittsburghTIGER2 locator. Save **OffenseUpdate20081101** to the chapter 9 geodatabase. The match statistics you should get are as follows:

Matched: 170 (90%)

Tied: 0 (0%)

Unmatched: 19 (10%)

Although it is possible to improve the match rate, you can skip that work in this exercise so that you can go on to new tasks.

Conduct spatial overlay of matched offense points

Spatial join, also known as spatial overlay, assigns polygon identifiers to points. In this exercise, you overlay police sector boundaries on offense points, so that each offense record now has a police sector. This makes it possible to produce monthly crime counts by sector. Although you are only adding a police sector identifier at this point, you could use the same set of steps to overlay a second, or even a third, set of boundaries on the offense points so that the master file is ready for data aggregation to additional polygon geographies, such as census tracts. Note that in the following steps, you process the master file as well as the update file, although in practice, the spatial join of the master file would already be completed.

1   In the table of contents, turn all layers off except PoliceSectors. Turn PoliceSectors on.

2   Right-click PoliceSectors and select Properties. Click the Labels tab and select the "Label features in this layer" check box. For Label Field, select Sector. Click OK. The police sector number has two parts: the police precinct and the patrol area. Police Sectors 4-1 through 4-7, for example, are the seven patrol areas that make up precinct 4. Clearly, the precinct number and the patrol area are two of the essential elements for data aggregation and reporting.

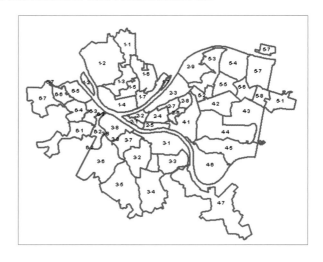

3 Open the ArcToolbox window and expand the Analysis toolbox and then the Overlay toolset. Then double-click the Spatial Join tool. **Warning:** It is crucial to use the Spatial Join tool for data aggregation, and not the Intersect tool or the Union tool. Using one of these will lead to duplicate records for points that are street intersections and are on the boundaries of the polygons used for data aggregation. Using the Intersect or Union tool creates a new point in the results for each polygon that shares the street intersection.

4 Make selections as shown in the figure. Save the Output Feature Class, **OffenseMaster20081031TempOverlay**, to the chapter 9 geodatabase in MyExercises.

5 Click OK.

6 In the table of contents, right-click OffenseMaster20081031TempOverlay and select Open Attribute Table. Scroll to the right to see that Sector has been added as an attribute. Then close the attribute table.

7 Remove OffenseMaster20081031c and OffenseMaster20081031Temp from the table of contents.

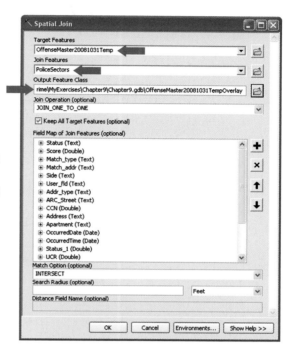

YOUR TURN **YOUR TURN** YOUR TURN YOUR TURN YOUR

Use the Spatial Join tool to overlay PoliceSectors on Geocoding Result: OffenseUpdate20081101 to produce **OffenseUpdate200811010verlay**. Then remove Geocoding Result: OffenseUpdate20081101 from the table of contents.

Calculate day-of-week supplemental field

For data processing and query applications of address-matched data, it is sometimes necessary to calculate new attributes (fields) by using the Field Calculator interface in ArcMap. One new field, WeekDay, is a code for the day of the week, where Sunday is 1, Monday is 2, Tuesday is 3, and so on. You can use any Visual Basic for Applications (VBA) functions or expressions in Field Calculator. Here, you use the DatePart("interval", [Date]) function. Some values for interval are "yyyy" for year, "q" for quarter, "ww" for week, and "w" for weekday.

1 In the table of contents, right-click OffenseMaster20081031TempOverlay and select Open Attribute Table. Click Table Options > Add Field. Type **WeekDay** for Name. Click OK.

2 Scroll to the right in the table, right-click the WeekDay field heading, and click Field Calculator.

3 **In the WeekDay = panel, type DatePart("w",[OccurredDate]).**

4 **Click OK.** If you checked a calendar, you'd see that the WeekDay field now has the correct codes for the day of the week. For example, 1/1/2008 was a Tuesday, WeekDay = 3. Now, you have the data for querying by the day of the week.

YOUR TURN **YOUR TURN** YOUR TURN YOUR TURN YO

Create the same attribute, **WeekDay**, in the attribute table for OffenseUpdate20081101Overlay.

Append update file to the master offense file

The next step is to append OffenseUpdate20081101Overlay to OffenseMaster20081031TempOverlay to produce the new master file, OffenseMaster20081101.

1 **In ArcToolbox, expand the Data Management toolbox and then the General toolset. Then double-click the Append tool.**

2 **Make selections as shown in the figure.**

3 **Click OK.**

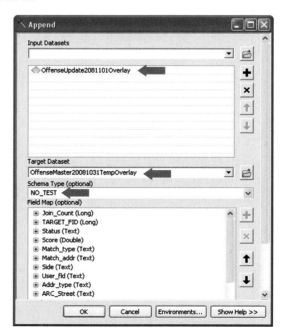

4 Open the attribute table of OffenseMaster20081031TempOverlay and sort OccurredDate by ascending order. Scroll to the bottom of the table and note that the update data for 11/1/2008 is appended. You should have 51,815 records in the table. If you do not have that number and wish to proceed without finding your error, remove your copy of OffenseMaster20081031TempOverlay from the table of contents, add OffenseMaster20081101 from the chapter 9 geodatabase in FinishedExercises, and skip steps 5–7. Otherwise, proceed as follows.

5 Close the table.

6 In the table of contents, right-click OffenseMaster20081031TempOverlay and click Remove. Then remove OffenseUpdate20081101Overlay.

7 In Catalog, rename OffenseMaster20081031TempOverlay in the chapter 9 geodatabase to **OffenseMaster20081101** and add it back to the table of contents.

8 In Catalog, delete all Offense files in the chapter 9 geodatabase, except OffenseMaster20081031, OffenseUpdate20081101, and OffenseMaster20081101.

9 Save your map document.

Tutorial 9-4

Aggregating data

Operational-level personnel, such as field officers, generally need information on individual events as they affect the direct delivery of services. Managers, however, often need information from aggregate data, such as the number of burglaries per month by police sector, which they use as input for decision making. Information from this kind of data can help managers answer questions such as, "Which police sectors had the highest increase in burglaries this month compared to last month?" The sectors that have the highest increase could be good candidates for additional police resources such as burglary squad detectives.

Select records to aggregate

Counts of incidents by polygon for a single time interval are easy to get with the use of ArcMap tools. In this exercise, you create counts of burglaries for January and then February 2008, and then append the results as the start of a space and time series dataset for burglaries.

1 Save your map document as **Tutorial9-4.mxd** to your chapter 9 folder in MyExercises.

2 From the Menu bar, click Selection > Select By Attributes.

The next step is long, but you can try to build the query command yourself by referring to the finished query in the following figure. Be sure to get the start and end dates correct for January 2008 by using the Get Unique Values button in the Select By Attributes dialog box.

3 In the Select By Attributes dialog box, select OffenseMaster20081101 for Layer. For the query expression, scroll down the list of attributes, double-click "Hierarchy", and click the = button. Type a space and **5** (to yield "Hierarchy" = 5). Then click And. Double-click "OccurredDate", click >=, and click Get Unique Values. Double-click date '2008-01-01 00:00:00'. Then click And. Finally, double-click "OccurredDate", click <=, scroll down the unique-values panel, and double-click date '2008-01-31 00:00:00'.

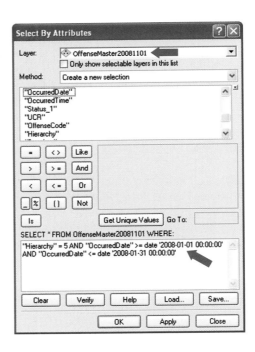

4 Click OK. The query selects all hierarchy 5 (burglary) crimes for January 2008.

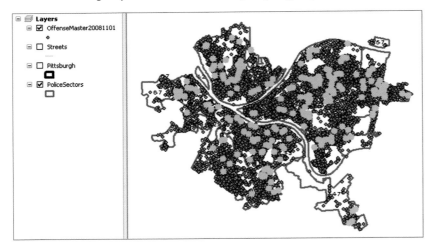

5 Right-click OffenseMaster20081101 and select Open Attribute Table. You should find 262 records selected for January 1–31, 2008.

Aggregate data in ArcMap

ArcMap has a built-in function that counts or otherwise summarizes selected records—in this case, it counts the records by police sector. So, the output has a record for each police sector and the number of burglaries in that sector for January 2008.

1 Scroll to the right, right-click the Sector column heading, and click Summarize. As with most functions in ArcMap, Summarize, by default, inputs only selected records if any are selected. Otherwise, ArcMap functions input all records.

2 In the Summarize dialog box, change the output table for item 3 to **Burglary200801** and save it to the chapter 9 geodatabase. Click OK > Yes.

3   Close the attribute table and open Burglary200801. It would be helpful to have the count of burglaries per month in a field called Burglaries, but ArcMap created a field called Cnt_Sector with the alias name Count_Sector displayed. ArcMap does not have the functionality to rename fields, so you'll have to create a new field called Burglaries. Then you can copy the contents of Cnt_Sector to Burglaries and delete the Count_Sector field from the table. The sector with OBJECTID = 1 corresponds to burglaries outside of Pittsburgh, which in this case is none.

4   Click Table Options > Add Field. Type **Burglaries** for Name and click OK.

OBJECTID *	Sector	Count_Sector
1		0
2	1-1	2
3	1-2	10
4	1-3	15
5	1-4	1
6	1-5	2

5   Right-click the Burglaries column heading and select Field Calculator. In the Fields panel, double-click Cnt_Sector. Click OK.

6   Right-click the Count_Sector column heading and click Delete Field. Click Yes.

7   Add two more fields, one called **Year** and the other called **Month**, and make both of them Short Integer.

8   Right-click the Year column heading and select Field Calculator. Clear the bottom Year = panel and type **2008**. Click OK.

9   Similarly, set Month to the value **1** and close the table.

OBJECTID *	Sector	Burglaries	Year	Month
1		0	2008	1
2	1-1	2	2008	1
3	1-2	10	2008	1
4	1-3	15	2008	1
5	1-4	1	2008	1
6	1-5	2	2008	1

YOUR TURN **YOUR TURN** YOUR TURN YOUR TURN YOUR

Open the attribute table for OffenseMaster1101 and repeat steps 1–9 to produce **Burglary200802** with burglaries by police sector for February 2008. Save it to the chapter 9 geodatabase. You should find that there were 165 burglaries in Pittsburgh that month. When you are finished, clear any selections. **Hint:** Carefully modify the dates in the current query to get the dates for February. Select "OccurredDate" in the current query and click Get Unique Values to select dates in the query and replace them with February dates. Be sure to create new fields that have exactly the same names and data types as the January query. Otherwise, the two aggregate tables will not append in the next exercise.

When you are finished, close any open tables and clear selected features. Save your map document.

Append table data in ArcMap

Space and time series data needs to have all the police sector data and all the monthly data in the same table, so now, you need to append data for January into the February table. Burglary200802 will thus become the master table for 2008 and include all the aggregate data through February.

1   In ArcToolbox, expand the Data Management toolbox and then the General toolset. Then double-click the Append tool.

2   Select Burglary200801 for Input Dataset, Burglary200802 for Target Dataset, and NO_TEST for Schema Type. Click OK. Then close ArcToolbox.

3   Open the Burglary200802 table and note that it now has data for both months, January and February.

4   Select both the Year and Month columns, right-click the column heading for either column, and select Sort Ascending. The space and time series data is now in the correct order. Next, you could aggregate data for March, although not done in these exercises, to create the table Burglary200803 and append Burglary200802 for a space and time series for all the aggregate data through March. This process could be repeated every month, so that the latest month is always the master file, or map layer, for all historical aggregated data.

5   When you are finished, close the table and save your map document.

Tutorial 9-5

Protecting privacy in location data

Many police departments post crime maps on the Web to keep the public informed of crime risks. Although these Web sites make the necessary information available to the public in an effort to prevent crimes, the individual privacy of victims and potential victims can still be protected in these crime maps. The approach used in this section is to "fuzz" x- and y-coordinates of crime incidents so that the exact crime locations are not used. Instead of plotting incidents as close as possible to the actual street address, you can calculate new coordinates to occur randomly within the same block. This process uses Excel spreadsheet software to add random numbers to block centroid coordinates. It is the approach the Washington, D.C., Metropolitan Police Department uses in its public crime maps to protect individual privacy.

Calculate polygon centroids

1   Open Tutorial9-5.mxd in ArcMap. This map document has the blocks polygon layer and the offenses layer, which has a definition query yielding all mapped offenses for July 2008. The block layer's attribute table does not have x- and y-coordinates for block centroids, but ArcMap has a function to add them. You'll try that out next.

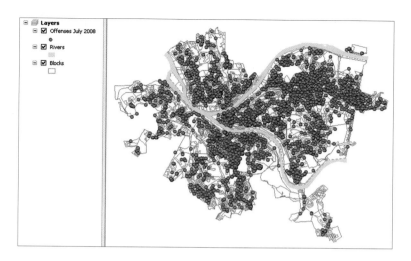

2   Save your map document to your chapter 9 folder in MyExercises.

3   In the table of contents, right-click Blocks and select Open Attribute Table.

4   Click Table Options > Add Field. Type **X** for Name and select Double for Type. Click OK.

5   Repeat step 4 to add the **Y** field.

6   Right-click the X column heading and select Calculate Geometry. Click Yes. For Property, select X Coordinate of Centroid. Click OK.

7   Repeat step 6 to compute the Y Coordinate of Centroid.

FID	Shape '	OBJECTID	BLKIDFP00	Shape_Leng	Shape_Area	X	Y
0	Polygon	1	420032507001029	946.199601	41055.393258	1337035.57158	418516.163244
1	Polygon	2	420032107002003	1627.614639	88197.0056	1335744.75882	417237.087807
2	Polygon	3	420032507001033	924.436852	42069.51039	1336748.90863	418315.409402
3	Polygon	4	420032507001030	550.128088	18246.792107	1337290.26169	418562.555745
4	Polygon	5	420032107002012	1100.189362	63867.763186	1336630.17208	417079.809194
5	Polygon	6	420032107002046	1405.710049	78135.671335	1335996.60371	415873.945166
6	Polygon	7	420032107002050	1758.992286	184047.343326	1336055.65065	415474.052982

8   Close the attribute table.

Overlay points with polygons

1   If ArcToolbox is not already available, click ArcToolbox on the Standard toolbar.

2   In ArcToolbox, expand the Analysis toolbox and then the Overlay toolset. Then double-click the Spatial Join tool.

3   Enter selections as shown in the figure and name the Output Feature Class **July2008Blocks**.

4   Click OK.

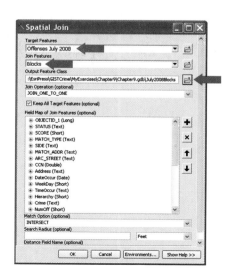

5   In the table of contents, right-click
    July2008Blocks and click Joins and Relates > Join.

The next step joins a code table to crime points so
that you have not only the hierarchy number for
identifying crimes, but also the name of the crime
type in the Crime attribute.

6   Enter selections as shown in the figure.

7   Click OK.

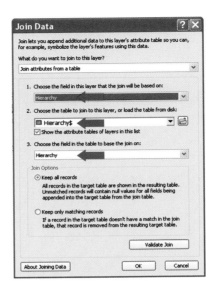

8   In the table of contents, right-click July2008Blocks and select Open Attribute Table. Then scroll to
    the right. The X and Y fields are block centroids. Also note the result of the join with the new
    Crime attribute.

July2008Blocks

BLKIDFP00	X	Y	Crime	Hierarchy	Part
▸ 420030402002004	1351756.596904	411607.263424	Simple Assault	10	2
420031302002011	1371615.241195	416853.912554	Drug Violations	18	2
420030603001014	1353109.801424	420198.384559	Robbery	3	1
420031915002031	1339142.067773	402027.470799	Drug Violations	18	2
420030511001001	1348513.900803	412068.957342	Drug Violations	18	2
420030201003003	1343914.567935	412769.513573	Drunken Driving	21	2

9   Click Table Options > Export. Click the Browse button and go to your chapter 9 folder in MyExercises.
    Change "Save as type" to Text File and type **OffensesJuly2008.txt** for Name. Click Save > OK > No.

10 Close the attribute table and save your map document.

In the next exercise, you'll use Excel to add random numbers to X and Y. This results in crime points
that are moved around enough in the same block so that you can still see separate point markers for
each point when zoomed in close enough to blocks.

Create random numbers

1   Start Excel. Click Office Button > Open. Change "Files of type" to All Files (*.*). Go to your chapter 9 folder
    in MyExercises (or the chapter 9 folder in FinishedExercises) and double-click OffensesJuly2008.txt.

2   In the Text Import Wizard step 1, make sure the Delimited option is selected and click Next. Then in
    step 2, select Comma for the delimiter and click Next. Finally, in step 3, click Finish.

3   Delete columns so that only the following columns remain: DateOccur, X, Y, and Crime_1. (You can
    delete several columns at once by dragging across the column headings to select them, and then right-
    click and click Delete.) Make the DateOccur field wider if necessary so you can see its values.

4 Change the name in cell D1 from Crime_1 to **Crime**.

5 Click in cell E1 and type **XRandom**. Then click in cell F1 and type **YRandom**.

	A	B	C	D
1	DateOccur	X	Y	Crime
2	7/1/2008 0:00	1351756.597	411607.2634	Simple Assault
3	7/1/2008 0:00	1371615.241	416853.9126	Drug Violations
4	7/1/2008 0:00	1353109.801	420198.3846	Robbery
5	7/1/2008 0:00	1339142.068	402027.4708	Drug Violations
6	7/1/2008 0:00	1348513.901	412068.9573	Drug Violations

6 Click in cell E2, type **=RANDBETWEEN(-50,50)**, and press the TAB key. Then click in cell F2 and type **=RANDBETWEEN(-50,50)**, and press the TAB key. The RANDBETWEEN function creates a uniform random number between -50 and 50 in each of the two cells.

7 Click in cell E2, drag across to cell F2 to select both cells, and press CTRL+C to copy the cell contents.

8 Click In cell E3, press SHIFT, scroll down and click cell F3442 (bottom of the table) to select the range from E3 through F3442, and release the SHIFT key. Press CTRL+V to copy the formula throughout the range.

9 Click the E column heading to select the entire column, right-click in the selection, and click Copy. Then right-click in the selection again and click Paste Special. Click the Values option. Click OK. The E column now stores the values displayed rather than their formulas.

10 Repeat step 9 for the F column.

Add random numbers to coordinates

In this exercise, you create new X and Y columns and calculate new values for them by adding the random numbers to the original coordinate values.

1 Change cell B1 from X to **XOriginal** and cell C1 from Y to **YOriginal**.

2 Right-click the B column heading and click Insert.

3 Repeat step 2 so that you have two blank columns, B and C, and then type **X** in cell B1 and **Y** in cell C1.

4 Drag across column headings B and C to select these columns, right-click in the selection, and click Format Cells. For Category, select General. Click OK.

5 Click in cell B2, type **=D2+G2**, and press the TAB key. Then click in cell C2, type **=E2+H2**, and press the TAB key.

6 Use steps 7 and 8 of the previous exercise to copy cells B2 and C2 throughout the B3 to C3442 range.

7 Drag across column headings B and C to select those columns, right-click in the range, and click Copy. Right-click again in the range, click Paste Special, and click the Values option. Click OK. This step replaces the formulas in these columns with their values. As a result, you can now delete the input columns used in creating columns B and C.

8 Delete columns so that the only remaining columns are DateOccur, X, Y, and Crime.

9 Click Office Button > Save As and go to Other Formats. Change the type to .csv. Click Save > Yes. Close Excel with no further changes.

Add XY Data and export as a shapefile

You can always display a data table that has appropriate x- and y-coordinates as a point map in ArcMap that uses the XY data format. A feature class is a better format than XY data, however, so in this exercise, you export the XY data as a feature class.

1   On the Standard toolbar, click Catalog.

2   In the Catalog tree, go to your chapter 9 folder in MyExercises and right-click OffensesJuly2008.csv. Click Create Feature Class > From XY Table.

3   Click Coordinate System of Input Coordinates and select Projected Coordinate Systems > State Plane > NAD 1983 (US Feet) > NAD 1983 StatePlane Pennsylvania South FIPS 3702 (US Feet). Click Add > OK.

4   In the Output area, click the Browse button and go to the chapter 9 geodatabase in MyExercises. For Name, type **July2008Protected**. Then click Save > OK.

5   On the Standard toolbar, click Add Data and add the new feature class, July2008Protected, to the chapter 9 geodatabase.

6   Turn all other point layers off and zoom in to see that the crime incident points of July2008Protected are all centered randomly within blocks. With the use of these random point locations, the actual addresses of crime incidents can be kept private to protect the identities of potential crime victims.

7   Save your map document and close ArcMap.

Geocode Pittsburgh 911 calls for service data

CAD data provides some unique crime indicators, such as shots fired and gun calls for service, which are not available from any other source. A limitation of CAD data, however, is that it is a product of citizen perception and is thus less accurate than data collected by trained police officers. In addition, citizens in high-crime areas may distort incidents—for example, they may claim guns are involved to get a higher priority and quicker response time. CAD data is, nonetheless, a unique source of information that is useful to investigators studying crime patterns.

CAD data is often collected by county agencies rather than municipal police departments. For example, the data in these assignments was provided by Chief Robert Full of the Allegheny County Department of Emergency Services. Hence, the address data entry procedures and personnel practices are different from those of the Pittsburgh Police Bureau and what is used in the exercises.

For this assignment and the next, build new geocoding methods for CAD data that are similar to but different from those you built for the offense data and that take these different methods of recording address data into account.

Create new geodatabase and map document

Create a new personal geodatabase called **Assignment9-1.gdb** and save it to your assignment 9-1 folder in MyAssignments.

Create a new map document called **Assignment9-1YourName.mxd** and save it to your assignment 9-1 folder. Use relative paths for your map document and use the state plane for southern Pennsylvania (NAD83 US Feet) for the data frame coordinate system. Add the following map layers and data:

+ Streets and Pittsburgh from the Pittsburgh geodatabase in the Data folder
+ CADWeapons20081231 spreadsheet from the RawData folder, which has calls for service code values, SHOTS and GUN (This data is valuable for a violent-crime squad or task force.)

Address match data

Build a new address locator called **PittsburghCAD** and save it to your assignment 9-1 geodatabase. Use US Address - Dual Ranges as the street style along with the settings used in tutorial 9-1 but with no alias table. Add **Location** as an input address field. (Click the Add button in that panel.) Add / as a connector for intersections as used in the CAD data.

Batch address match the addresses in the CADWeapons20081231 spreadsheet. Save the results as **CADWeapons20081231** to your assignment 9-1 geodatabase.

Provide a table with match statistics. Include the following:

+ A breakdown of nonmatches into (a) blank or missing addresses and (b) nonstreet and nonintersection addresses
+ A further breakdown of nonmatches that have street addresses and intersection formats and values, with totals and percentages for nonmatches that are streets versus nonmatches that are intersections

Hint on statistics: Fill in the following table.

Status	Street Addresses	Intersections	Other	Total Locations
Matched				
Tied				
Unmatched				
Total				

There are a few complicated cases in the source data that you will ignore in getting statistics. To get statistics for Intersections, query for locations that contain the connector "/" and add a condition for Status as needed. Do not make this query any more complicated than suggested. For Street Addresses, query for locations that start with numbers between 1 and 99999 and are not intersections (do not contain the connector "/"). Then add a condition for Status. Finally, for Other, get statistics by subtraction of Total Locations - (Street Addresses — Intersections). Note that rows add up to the Total row. Save your query expressions for Total Street Addresses and for Total Intersections in a Word document, using copy and paste from the Select By Attributes interface.

Create alias table

Find five place names and their addresses. Create an alias table called **PittsburghCADAlias.csv** and save it to your assignment 9-1 folder.

Hint on aliases: "Arena" is short for "Mellon Arena." If you are unsuccessful in finding a bar, try a hotel with the same name.

Test your addresses, using the Find tool and your new address locator.

Do not rematch the CAD data in this assignment.

Create map

Create a map called **Assignment9-1YourName.mxd** for July 2008 shots fired and gun calls for service. Save it to your assignment 9-1 folder and use the following steps:

1   Right-click Geocoding Result: CADWeapons20081231, select Properties, and click the Definition Query tab.

2   Use Query Builder to build the following expression for July 2008 calls for service:

```
"EntryDate" >= date '2008-07-01' AND "EntryDate" <= date '2008-07-31'
```

3   Click the Symbology tab.

4   In the Show panel, click Categories > Unique Values.

5   Select CallType in the values field and click Add All Values. Click OK.

6   Click File > Export Map. Export a JPEG map image called **Map9-1.jpg** with 300 dpi resolution to your assignment 9-1 folder. There are shots fired and gun calls for service patterns in the resulting map that are interesting to police officers or researchers who are familiar with Pittsburgh.

What to turn in

Use a compression program to compress and save your assignment 9-1 folder to **Assignment9-1YourName.zip**. Turn in your compressed file.

Alternatively, or in addition as required, include the following in a Word document, **Assignment9-1YourName.docx**:

   ✦   Map9-1YourName.jpg, including a legend for CAD calls

   ✦   Match statistics table

   ✦   Query expression code for Total Street Addresses and Total Intersections

   ✦   Alias table

Assignment 9-2

Build space and time series data for 911 calls

If you have not worked assignment 9-1, read the introductory paragraphs there to get an overview of CAD data and its uses in crime mapping.

For this assignment, geocode CAD data on illicit drug dealing in Pittsburgh, aggregate data to census tract polygons, and produce a choropleth map that has drug counts by tract for a given month.

Create new map document

Create a new personal geodatabase called **Assignment9-2.gdb** and save it to your assignment 9-2 folder in MyAssignments.

Create a new map document called **Assignment9-2YourName.mxd** and save it to your assignment 9-2 folder. Use relative paths for your map document and use state plane for southern Pennsylvania (NAD83 US Feet) for the data frame coordinate system. Add the following map layers and data:

+ Streets, Tracts, and Pittsburgh from the Pittsburgh geodatabase in the Data folder (Of course, Streets is the TIGER reference data for geocoding.)
+ CADDrugs20081231 spreadsheet, which has illicit drug calls for service, from the RawData folder in Data

Address match data

If you did assignment 9-1, use your existing address locator, PittsburghCAD. Otherwise, build a new address locator called **PittsburghCAD** and save it to your assignment 9-2 geodatabase. Use US Address– Dual Ranges as the street style along with the settings used in tutorial 9-2 but with no alias table. Add **Location** as an input address field. Add / as a connector for intersections as used in the CAD data.

Batch address match the CADDrugs20081231 spreadsheet. Save the results as **CADDrugs20081231** to your assignment 9-2 geodatabase.

Aggregate address-matched CAD data

Create monthly counts of drug calls by tract for January, February, and March 2008, resulting in a master table called **CADDrugs200803** in a database file format and include all three months of data in a single column. Include the attributes **Drugs** (= count attribute created by Summarize function), **Year** (value 2008), and **Month** (values 1, 2, and 3 for month number). Save all outputs and files to the assignment 9-2 geodatabase.

Create map

Create a map for February 2008 drug calls for service by tract called **Assignment9-2YourName.mxd**. Save it to your assignment 9-2 folder and use the following steps:

1   Join **CADDrugs200803** to Tracts.

2   Create a definition query for Tracts as follows:

        CADDrugs200803.Year = 2008 AND CADDrugs200803.Month = 2.

3   Symbolize Drugs, using five quantiles and a monochromatic color ramp.

4   Create a map layout with title, map, and legend. Export the layout image as **Map9-2.jpg** to your assignment 9-2 folder.

5   Create a Word document called **Assignment9-2YourName.docx** and save it to your assignment 9-2 folder. Include geocoding match statistics for CADDrugs20081231 and the Map9-2 JPEG image.

What to turn in

Use a compression program to compress and save your assignment 9-2 folder to **Assignment9-2YourName.zip**. Turn in your compressed file.

OBJECTIVES

Learn the components of ModelBuilder models
Explore a completed model
Build a model to process crime incident data into master files
Build models to produce crime maps from master files

Chapter 10

Automating crime maps

The ModelBuilder application in ArcMap allows you to create macros, which are custom computer programs, by dragging icons into a flowchart and filling out forms. These macros, called models in ModelBuilder, link series of steps to produce final outputs as easily as the click of a button. Using models not only saves much time and tedious work, but it also yields crime maps that are accurate and consistent. In this chapter, you build a ModelBuilder model to automate much of the daily spatial processing of police report data to get it ready for mapping. Then you build additional models to automate the full-scale production of crime maps.

ModelBuilder concepts

Much of the work in a crime mapping and analysis system is repetitive. Every day, every shift, new police reports come in that need to be processed from tabular data into master map layers, and these layers need to be applied to produce maps of various kinds for various audiences. Although it is possible to do all this, as in the previous chapters, the work is long, tedious, and prone to error. ArcGIS has a macro facility called ModelBuilder that can automate much of this work, leaving crime analysts more time for creative applications and problem solving through the use of GIS tools and technology.

As an introduction to ModelBuilder, you use and review a finished model in ArcMap in the beginning exercises of this chapter. In subsequent exercises, you learn how to build two major models, and then in the assignments at the end, you build two more.

Models

Computer programs consist of a series of input and output transformations. Each transformation is a set of steps, or an algorithm, that takes input data and uses it to produce output data, or new output information. The programs in this chapter use input data to either do sophisticated geoprocessing or produce detailed maps.

ModelBuilder provides prewritten algorithms in the form of tools that are available in ArcToolbox. Each tool has properties you can change to modify its behavior, as well as a form with fields you can fill out to identify input and output data. Each tool performs a process by using inputs and producing outputs. A ModelBuilder model is made up of one or more processes that execute in sequence to produce intermediate and final outputs. For example, the input of one model is a crime master feature class, the intermediate product is a specific crime map layer queried from the crime master feature class, and the final output is a pin map symbolized for field officers.

User interface

Models generally have parameters, or variables the user can change, to meet the need at hand. For example, model parameters may include the crime type to be mapped (e.g., burglary or rape) and the start and end dates of the crimes reported in the map. Parameters thus generalize models, making them reusable by allowing the user to simply change the values of variables rather than requiring additional model building or programming. Changing the parameters of the model, and then running the model to get the results of geoprocessing is done through the user interface. In ModelBuilder, as in many application programs, the user interface is a computer form, which allows you to select or type values for the parameters that are used to control the model. The user interface thus allows a crime analyst, or even a clerk, to carry out complex GIS processing simply and quickly, much like driving a car by putting it in gear and pressing the pedal rather than having to go under the hood and work on the engine.

Documentation

A model must provide not only the functionality of processing, but also the documentation users need to run the model. The ModelBuilder flowchart is in itself a major part of the documentation in that it graphically displays the processes being used and provides user access to the tools needed for geoprocessing. In addition, users can write a model description that becomes a property of the model, ensuring that models can easily be used and reused by crime analysts and others.

Debugging

Finally, models, like all computer programs, have a tendency to include errors (or "bugs") during the design phase. It is easy for the model builder to make mistakes when selecting and implementing the tools to be used in a model process. Errors can also come about when model users apply a model. ArcMap provides error codes and other forms of feedback that can help you diagnose model bugs, find the problem, and then debug models for continued use.

Using ModelBuilder for automation

To get started, in tutorial 10-1, you explore a simple completed model to get an overview of ModelBuilder and the nature of models it produces. The model in this exercise allows you to specify a date range for the mapped property crime data and produces the pin map you symbolize in chapter 3 for public use. The exercise also provides some experience in debugging a model. You put bugs in the model, and then attempt to run it to see what kind of error messages you get.

After being introduced to a finished model, you learn how to build a model for spatial data processing in tutorial 10-2. The model you build geocodes updated crime data, uses spatial overlay to assign police sector IDs to crime points, and appends an update feature class to the master feature class to produce a new updated master feature class. The resulting model makes daily updates of crime data easy enough for a clerk to run. Without a model, the series of interactive steps to produce the same output are complex, requiring much expertise in the use of ArcMap, and are particularly prone to error because of the many detailed tasks and need for strict consistency.

Finally, in tutorial 10-3, you learn how to build a sophisticated model that produces the field officers' pin map you use in chapters 3 and 4. This model includes the ability to process data for temporal context through the use of brighter and larger point markers to represent more recent crimes. In the two assignments at the end of the chapter, you implement two other significant models. Assignment 10-1 is to implement a model for producing choropleth maps, and assignment 10-2 is to implement a model for producing a pin map that has size-graduated point markers for repeat crime incidents in the same locations.

Tutorial 10-1

Exploring a completed model

In this exercise, you learn how to run a model in different modes, examine the elements and properties of a model, review model documentation, and create and fix model bugs. Note that the master feature class used has data for only the month of July 2008, so you will need to restrict outputs to dates that fall within that month.

Add toolbox and model to ArcToolbox

1   Open Tutorial10-1.mxd in ArcMap.

2   Save your map document to your chapter 10 folder in MyExercises.

3   On the Standard toolbar, click the ArcToolbox window button. Then right-click the ArcToolbox icon and select Add Toolbox. Go to your MyTools folder, click the Crime Mapping Tools Starting toolbox, and click Open. The corresponding toolbox appears in the ArcToolbox window.

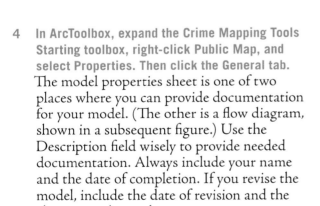

4   In ArcToolbox, expand the Crime Mapping Tools Starting toolbox, right-click Public Map, and select Properties. Then click the General tab. The model properties sheet is one of two places where you can provide documentation for your model. (The other is a flow diagram, shown in a subsequent figure.) Use the Description field wisely to provide needed documentation. Always include your name and the date of completion. If you revise the model, include the date of revision and the changes you've made.

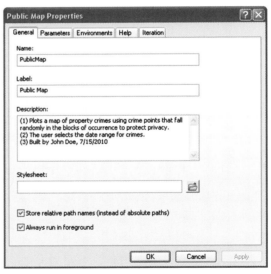

5   Click OK.

Run Public Map model

1   In ArcToolbox, right-click Public Map and click Open. The form that appears when you open a model is the ModelBuilder user interface for running a model. It includes the parameters you can change and places to view the Help documentation. You can change any of the parameters by typing or making selections from lists, if available. The form provides default start and end dates, which are in the needed date format, and the default name of the output crime map

layer—Property Offenses, as shown in the figure. The query expression is not meant for the user to change but is part of the permanent program.

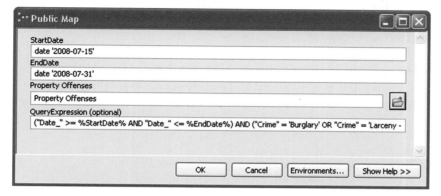

You can run the model with the default parameter values, although you can change the parameters as needed to run the model under different conditions.

2   **Click OK to run the model.** A runtime window opens that provides log information on the execution of the model as it runs. Running the model produces outputs for the date range.

3   **Click Close.** The model selects the desired crime types and date ranges from the source crime master file and renders the map with a legend.

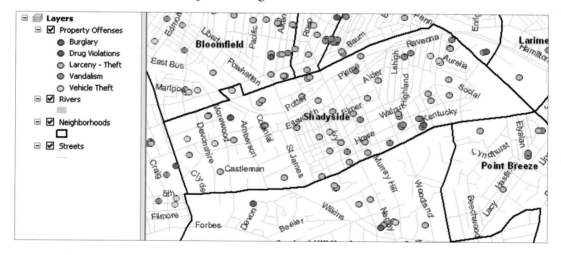

4   In the table of contents, right-click Property Offenses and click Remove so you can run the model again.

5   In ArcToolbox, right-click Public Map, click Open, and ignore the warning message. Change StartDate to **date '2008-07-01'** and change Property Offenses to **Property Offenses: July 2008**. Click OK > Close. The map now has more points displayed, given the new and larger date range, and the label in the table of contents for the output crime points layer now has the month and year of the output, July 2008.

6   In the table of contents, right-click Property Offenses: July 2008 and select Open Attribute Table. Right-click the DateOccur column heading and select Sort Ascending. The full date range of points, from July 1 to July 31, 2008, is now in the table, and therefore mapped.

7   Close the table and remove Property Offenses: July 2008 from the table of contents. Save your map document.

Examine model in Edit mode

1 **In ArcToolbox, right-click Public Map and select Edit.** The interface displayed is a model flow diagram for building and editing a model. The model diagram also serves as documentation through its flowchart structure and through the comments that you add for explanation. This model is simple with a single process, represented by the yellow rectangle called Make Feature Layer. The model creates the output map layer, Property Offenses, for display in the map document by applying query criteria to the July2008Protected feature class. The key element is QueryExpression, which incorporates the parameter values for StartDate and EndDate supplied as query criteria by the user.

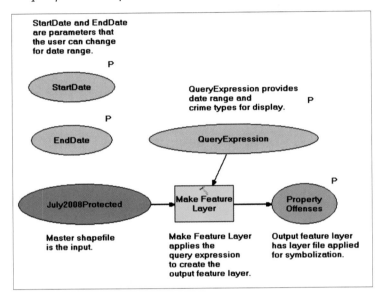

2 **Right-click StartDate and click Open.** This form is where you can enter a default value for the parameter. The value has to be in the format shown in the figure.

3 **Click Cancel.**

4 **Right-click Make Feature Layer, click Open, and ignore any error message.** This is the form interface for the Make Feature Layer tool. The input features, July2008Protected, are already selected. Note that QueryExpression is selected as the SQL code to create the output layer.

5 **Click Cancel.**

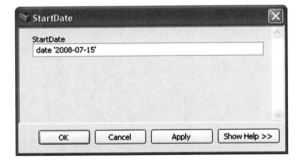

6 **Right-click QueryExpression and click Open.** The query expression that follows may look complicated, but it can be built by using a query builder for assistance, and then inserted into the model by copy and paste.

```
"DateOccur" >= %StartDate% AND "DateOccur" <= %EndDate%) AND
("Crime" = 'Burglary' OR "Crime" = 'Larceny - Theft' OR "Crime" = 'Vehicle
Theft' OR "Crime" = 'Vandalism' OR "Crime" = 'Drug Violations')
```

The notation %StartDate% uses % as the delimiter to indicate the need for the model to substitute the user's parameter value stored in StartDate in the query expression. %EndDate% works the same way. This programming step is called "in-line substitution." The query has two conditions that both must be TRUE, each enclosed in parentheses and connected by AND. The first condition selects only the records that are within the desired date range. The second condition selects only the records of the specified crime types (i.e., Burglary, Larceny–Theft, and so on). The second condition uses OR connectors, which need to be used to select miscellaneous subsets of larger collections.

7 **Click Cancel.**

YOUR TURN **YOUR TURN** YOUR TURN YOUR TURN YOUR

Open and study the remaining model elements.

Create and fix bugs in Edit mode

1 **Right-click and open QueryExpression.**

2 **In the QueryExpression field, type ? at the end of the first occurrence of DateOccur to produce "DateOccur?".** This expression creates a bug, because there is no attribute named "DateOccur?" in the attribute table being queried.

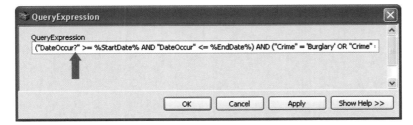

3 **Click OK.**

4 **From the Public Map model Menu bar, click Model > Run Entire Model.** The Public Map Run window indicates success and adds a legend for Property Offenses in the table of contents. However, no crime points are displayed in the map, and you get an ArcMap Drawing Errors message window.

5 **Close the Public Map Run window.** The ArcMap Drawing Errors message that a field was not found is helpful. It is exactly the kind of bug you created. But if you were to get this message while trying to run a model, you'd have to use this clue to search for the bug on your own. Hopefully, in the event of a real bug, you would know to study your query condition for the problem. If necessary, you could open the attribute table of the input feature class to note the correct spellings of field names, and perhaps detect the bug that way.

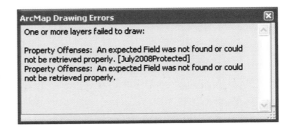

6 **Close the ArcMap Drawing Errors window and open QueryExpression. Fix the bug by deleting ?** in **"DateOccur?". Click OK. Then remove Property Offenses from the table of contents.** If you were to run the model again, you'd know that it would execute correctly. Instead, in the next steps, create a second bug.

7 **Right-click StartDate and click Open.**

8 **Create a bug by deleting the final single quotation mark in the value for StartDate, yielding** date '2008-07-15'. **Click OK.**

9 **From the Public Map model Menu bar, click Model > Run Entire Model. Then click Close.** This time, the message in ArcMap Drawing Errors—that an invalid SQL statement was used—is less useful than in the previous case. You should be suspicious that QueryExpression has an error, but the problem is all the way back at StartDate, which has a date value in an incomplete format.

10 **Close the ArcMap Drawing Errors window. Fix the date in StartDate by adding the single quotation mark at the end of the query expression. Click OK and save the model. Remove Property Offenses from the table of contents.**

YOUR TURN YOUR **TURN** YOUR TURN YOUR TURN YOU

So far, you have looked at bugs, or errors in the code that prevent outputs. Another kind of error is a logic error—in which case, the model runs without a problem but fails to produce the expected results. In this exercise, create a logic error. Open QueryExpression and change <= in <= %EndDate% to yield >= %EndDate% so that that part of the query criteria is ("DateOccur" >= %StartDate% AND "DateOccur" >= %EndDate%). What is the only date that satisfies that criterion? Run the model with the error and see. Take a look at the Property Offenses attribute table. Then clean things up by removing Property Offenses from the table of contents and fix the error. When you are finished, save the model and close the Public Map model window. Then save your map.

Tutorial 10-2

Processing police reports into master files

Now that you've seen how a model works, you can create a model to automate the geocoding and spatial data processing done in chapter 9. You start at the beginning by opening a new map document, setting the properties for ModelBuilder, and creating a new toolbox and tool for crime mapping. Then you learn the steps for building, testing, and documenting the processes of the model.

Set properties for ModelBuilder

1   Open Tutorial10-2.mxd in ArcMap. Save your map document to your chapter 10 folder in MyExercises. The map document opens with the layers PoliceSectors, Rivers, and Streets. The next step sets the properties for geoprocessing with the use of ModelBuilder.

2   From the Menu bar, click Geoprocessing > Geoprocessing Options and enter selections as shown in the figure if they are not already selected. Because of the first option you set, "Overwrite the outputs of geoprocessing options," ArcMap will overwrite previous versions of outputs when you rerun the model, thus preventing a proliferation of files. The last option you set, "Add results of geoprocessing operation to the display," makes ModelBuilder automatically add new map layers to your map document when a model is run.

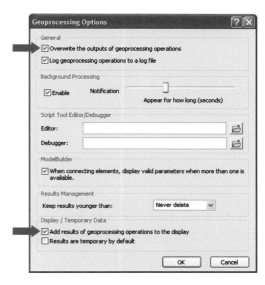

3   Click OK.

4   On the Standard toolbar, click ArcToolbox. In the ArcToolbox window, right-click the ArcToolbox icon and select Environments.

5   In the Environment Settings window, click the Workspace line and enter selections as shown in the figure.

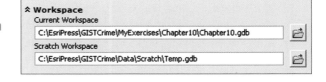

6   Click OK.

Create new toolbox

1   From the Menu bar, click Windows > Catalog.

2   Expand the Catalog tree to your chapter 10 folder in MyExercises and right-click the chapter 10 folder icon. Click New > Toolbox.

3   Rename the toolbox **CrimeMappingTools.tbx**.

4   In ArcToolbox, right-click the ArcToolbox icon and select Add Toolbox. Go to your chapter 10 folder in MyExercises and click the CrimeMappingTools toolbox. Click Open.

5   If necessary, close the Catalog window.

Create new model

1   In ArcToolbox, right-click the CrimeMappingTools toolbox and select New > Model.

2   Close the resulting model window. Then in the ArcToolbox window, right-click Model and rename it **Process Police Reports**.

3   Right-click Process Police Reports and select Properties. Click the General tab and enter selections as shown in the figure.

4   Click OK.

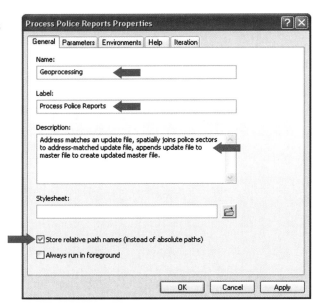

Build "geocoding addresses" process

The first step in processing new police report data is to extract update data from the police records management system. This is the data that is new to the crime mapping and analysis system, and it starts at exactly the point where the existing master map layer leaves off. The person extracting this data needs to carefully check the end date and time of the current master files and the start date and time of the update files to ensure that the data from both files matches perfectly.

1   In ArcToolbox, right-click Process Police Reports, select Edit, and resize the resulting model window to make it about 50% wider and taller.

2   In ArcToolbox, expand the Geocoding toolbox and drag the Geocode Addresses tool into the upper-left corner of the Process Police Reports model window. Then move the elements so there is about 2 inches of space to the left and above them.

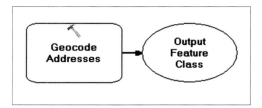

3   Click anywhere in the white area of the model window to clear the selection of the elements and double-click Geocode Addresses.

4 Click the Browse button for each of the parameters of this process to fill in Input Table, Input Address Locator, and Output Feature Class as shown in the figure.

5 Click OK.

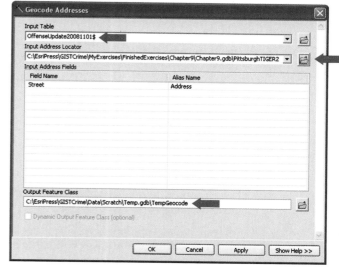

6 Change the size of the elements by dragging the "handles" to improve readability.

7 Save the model.

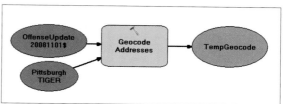

Build spatial join process

Now, you can assign police sector IDs to each geocoded point in TempGeocode. This process prepares the data for aggregation to crime counts by police sector and time period.

1 In ArcToolbox, expand the Analysis toolbox and then the Overlay Toolset. Then drag the Spatial Join tool to the right of TempGeocode in the model window.

2 In the model window, double-click Spatial Join, and then enter selections as shown in the figure.

3 Click OK.

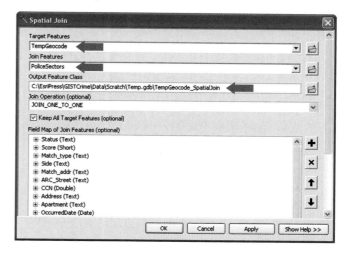

4 Rearrange and resize elements to make the model more readable and save the model.

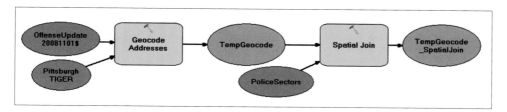

Build copy process

You can now append the update file to the current master file to create a new master file. To preserve the current master file, in case processing fails or there is a mistake, it is best to make a copy as the basis for the updated version. Then if disaster strikes, you can repeat the steps and use the original copy of the current master file.

1   In ArcToolbox, expand the Data Management toolbox and then the General toolset. Then drag the Copy tool below the Geocode Addresses process in the model window.

2   Double-click the Copy element and enter selections as shown in the figure. When browsing for "Output data element," double-click the chapter 10 geodatabase, change the Save As type to Datasets, and then type **OffenseMaster20081101** for Name.

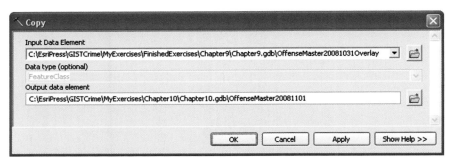

3   Click OK, rearrange or resize elements for readability, and save the model.

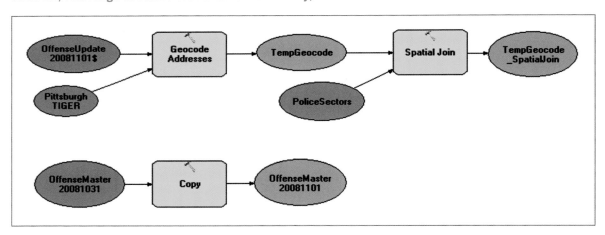

Build append process

1   In ArcToolbox, expand the Data Management toolbox and then the General toolset. Then drag the Append tool to the right of OffenseMaster20081101 in the model window.

2   Double-click the Append element and enter
    selections as shown in the figure.

3   Click OK. Save the model.

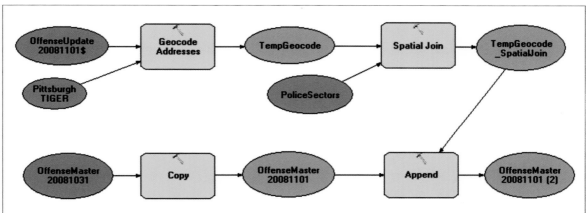

This completes building the model. Note that, in general, it is necessary to aggregate the data in the resulting master shapefile to monthly crime counts by police sector. ArcToolbox does not have the tools for this step, but the work can be done by using ArcMap interactively to process the new master file attribute table (see chapter 9).

Set parameters

The user will want to change the values of parameters when running this model—in particular, the input update dataset (Updated Data Table), the input master file (Old Master Feature Class), and the name of the output master file (New Master Feature Class Completed). All you have to do when designing the model is designate the desired inputs and outputs as parameters so the interface will allow these changes.

1   In the model window, right-click the OffenseUpdate20081101$ element and select Model Parameter. ModelBuilder adds a *P* to the display near the OffenseUpdate20081101$ element to indicate that it is a parameter, and thus changeable by the user.

2   Repeat step 1 for the input and output of the copy process, OffenseMaster20081031Overlay and OffenseMaster20081101, respectively.

3 Right-click the output of the append process, OffenseMaster20081101 (2), and select Add To Display.

4 Right-click TempGeocode and then TempGeocode_SpatialJoin.

Notice that they are designated Intermediate, which means that ArcMap deletes them from your data files after running the entire model as long as the model is run from the user interface. In the next exercise, you'll use the interface to run the model.

Run model

When you are debugging the model, it is best to run it in Edit mode, which opens the model window, so that you can make changes quickly and easily.

1 Save your model.

2 From the Menu bar of the Process Police Reports model window, click Model > Run Entire Model. If your model has no errors, it should run smoothly and produce the desired results.

3 In the table of contents, open the attribute table of the output of the append process (New Master Feature Class), sort by OccurredDate, and verify that there are 51,815 records, including the added records of 11/01/2008.

4 Close the table and remove OffenseMaster20081101 from the table of contents.

5 Save and close the model window.

6 Right-click your Process Police Reports model in the ArcToolbox window and click Open. The warning message is just there to say that the OffenseMaster20081101 feature class already exists.

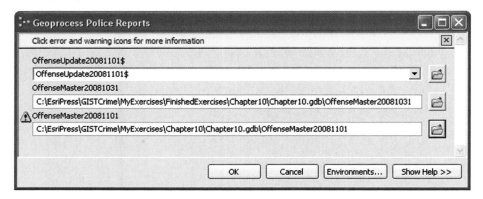

7 Click OK. The model runs again, overwriting the previous output, OffenseMaster20081101, with the new output. It is another successful run of the model, this time from the end user's perspective.

8 Save and close your model.

Generalize model element names

You can use the same model to update *any* master point feature class—offense data, CAD data, or any other data that has street addresses. All that is needed to generalize the model is to add more parameters, generalize model element names, and add self-documenting labels.

1 In ArcToolbox, right-click Process Police Reports and select Edit.

2 Make PittsburghTIGER and PoliceSectors parameters. By making these elements parameters, you can use any street map or any other reference data in your model and use any polygons for spatial joins.

3 Right-click selected elements and variables and select Rename. Change the names as shown in the figure, and rearrange and resize the elements for readability.

4 Make New Master Feature Class Completed a model parameter.

5 Save and close your model.

6 Save your map document.

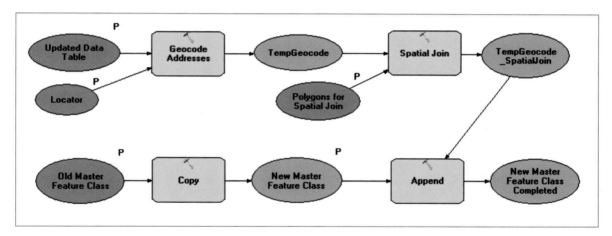

Tutorial 10-3

Producing a pin map for field officers

With an updated master feature class ready for use, the next task is to produce the field officers' pin map, based on symbolization layer files created in chapter 4. In this exercise, you use the updated master feature class provided in the chapter 10 folder in FinishedExercises.

Set properties for ModelBuilder

1   Open Tutorial10-3.mxd in ArcMap. The map document opens with the familiar layers: PoliceSectors, Rivers, and Streets.

2   Save your map document to your chapter 10 folder in MyExercises.

3   From the ArcMap Menu bar, click Geoprocessing > Geoprocessing Options. Verify that the following two options are selected: "Overwrite the outputs of Geoprocessing operations" and "Add results of Geoprocessing operations to the display."

4   Click OK.

5   On the Standard toolbar, click ArcToolbox. Then in the ArcToolbox window, right-click the ArcToolbox icon and select Environments.

6   Click the Workspace line and enter selections as shown in the figure.

7   Click OK.

Create new model

1   In the ArcToolbox window, right-click the ArcToolbox icon and select Add Toolbox. Go to your chapter 10 folder in MyExercises and click the CrimeMappingTools toolbox. Click Open.

2   In ArcToolbox, right-click CrimeMappingTools and select New > Model. **Note:** A model must be created within a toolbox.

3   Close the resulting model window. Then, in ArcToolbox, right-click Model and rename it **Produce Field Officers' Pin Map**.

4   In ArcToolbox, right-click Produce Field Officers' Pin Map, select Properties, and click the General tab. Enter selections as shown in the figure.

5   Click OK.

Create variables

This model uses date arithmetic to create a data window, or period of time in days, using an end date and the total length in days as inputs. Date arithmetic saves the user the effort of using a calendar to figure out a date in the window to separate and define old data from recent data. The basis for calculations is the master file attribute DataAge. The new model calculates values for DataAge as EndDate = 0, where EndDate is the end date of the data window, and older dates as the number of days prior to the end date, or EndDate - OccurredDate. ModelBuilder uses the VBScript scripting language, a small subset of the Microsoft Visual Basic computer language, for programming syntax.

1   In ArcToolbox, right-click Produce Field Officers' Pin Map, select Edit, and make the window about 50% larger.

2   Right-click in the upper-left corner, click Create Variable, and select Date for Data Type. Click OK.

3   Right-click Date, click Rename, and type **EndDate** for New Element Name. Click OK.

4   Right-click EndDate, click Open, and type **7/11/2008** for EndDate Default Value. Click OK. When the model has expressions that are correct and can locate the needed inputs, it turns expression ellipses blue to indicate that they are correct and ready for use.

YOUR TURN   **YOUR TURN**   YOUR TURN   YOUR TURN   YOUR

Create four more variables. Two of the variables have Variant for Data Type: **TotalWindow** with value 28 and **RecentWindow** with value 7. (Variant is a generic data type in the VBScript scripting language that underlies some of the processing in ModelBuilder. VBScript senses the type of data needed for variant variables, such as integer in this case, by context or through entered values.) Create a third additional variable called **DataAgeExpression** with String for Data Type. String is another name for text. You'll use this variable in the next exercise to create an algebraic expression that takes the value of EndDate and subtracts the values of the field OccurredDate (incident occurred date). It does not get a default value. Finally, create another String variable called **OldIncidentExpression**. It does not get a default value either.

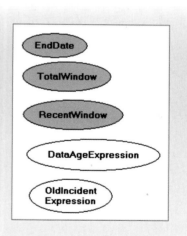

Build "calculate field" process

Computers store dates using a Julian calendar, in which a very old start date is given the number 1 and each successive day in the calendar gets a sequence number added to the start date. When you subtract two dates on a computer, the result is thus the number of days between the dates (or a time interval in some other unit such as hours). In VBScript expressions, date constants must be surrounded by pound signs (for example, #7/15/2008#) so that VBScript can distinguish date constants from variables. As you saw earlier in this chapter, ModelBuilder signals that it is using the value from another variable by placing the name of that variable in percent symbols (for example, %EndDate%). Finally, ModelBuilder expressions must indicate field names from attribute tables by placing them in square brackets (for example, [OccurredDate]).

Hint: The following steps require detailed typing. If you want to avoid typographical errors, you can find the needed expressions in the Expressions text file in the chapter 10 folder in FinishedExercises. Simply open this file in Notepad, copy the expressions one at a time, and paste them in the appropriate spot where extensive typing is required.

1   Right-click DateAgeExpression, click Open, and type the following value:

          #%EndDate%# - [OccurredDate]

Click OK. This expression calculates the age of each incident in days relative to EndDate and is the basis for creating a data window. If OccurredDate is the same day as EndDate, DataAge is 0, so for a 28-day window, DataAge ranges between 0 and 27 (instead of between 1 and 28). In regard to syntax, EndDate is a variable, and so needs to be enclosed in percent symbols (%) to indicate to ModelBuilder to get its value from that variable. Furthermore, EndDate needs to be enclosed in pound signs (#) because it is a VBScript date value. OccurredDate is an attribute from the OffenseMaster20081101 attribute table, and thus is designated by square brackets ([]).

2   In ArcToolbox, expand the Data Management toolbox and then the Fields toolset. Then drag the Calculate Field tool inside the model window.

3   Right-click the Calculate Field element and click Open.

4   Enter selections in the Calculate Field dialog box as shown in the figure.

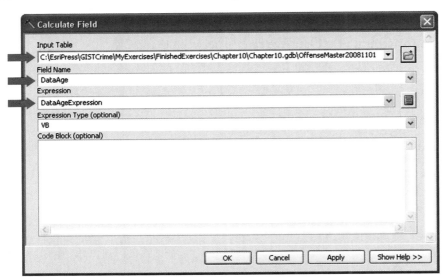

5   Click OK.

6   In the model window, rearrange and resize elements to improve readability and appearance.

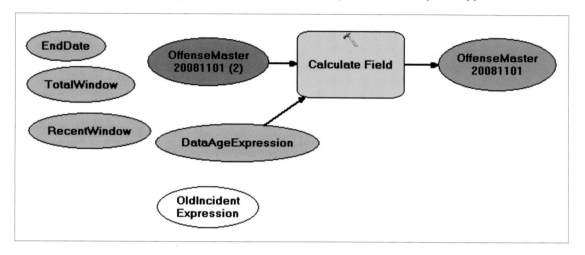

Build "make feature layer" process

In this exercise, you create a feature layer that extracts desired data from the master map file. This layer is for the "old" data for Part 1 crimes, which extracts data for the first three weeks of the data window. A feature layer simply points to an existing feature class (OffenseMaster20081101, in this case), extracts desired features from it, and displays it with an entry in the table of contents.

1   Right-click OldIncidentExpression, click Open, and type the following value on a single line, although it is shown on four lines here for ease of reading and interpretation. (Recall that you can copy and paste this expression from the Expressions text file in the chapter 10 folder in FinishedExercises.)

```
"DataAge" >= %RecentWindow% AND
"DataAge" <= (%TotalWindow%-1) AND
"Hierarchy" >= 1 AND
"Hierarchy"
```

Click OK. This logical expression retrieves data that is at least as old as RecentWindow (7 days) and as old as TotalWindow–1 (27 days) and includes all serious or Part 1 crimes (hierarchy 1–7). The DataAge range takes into account that DataAge ranges between 0 and 27.

2   In ArcToolbox, expand the Data Management toolbox and then the Layers and Table Views toolset. Then drag the Make Feature Layer tool into the model window under the Calculate Field element.

3   On the ModelBuilder Standard toolbar, click the Connect tool ⚡, and then click OldIncidentExpression > Make Feature Layer > Expression.

4   Click OffenseMaster20081101 > Make Feature Layer > Input Features.

5   Click the Select button ▸. After using a different ModelBuilder tool, such as the Connect tool, always go back to the Select tool as done in step 5.

6   Right-click Make Feature Layer, click Open, and type the Output Layer name as shown in the figure.

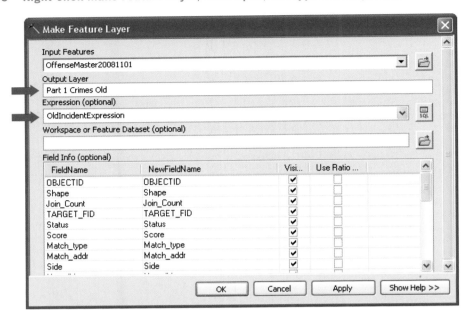

7   Click OK.

8   Right-click Part 1 Crimes Old and select Model Parameter. Making Part 1 Crimes Old a parameter allows you to rename the output when running the model.

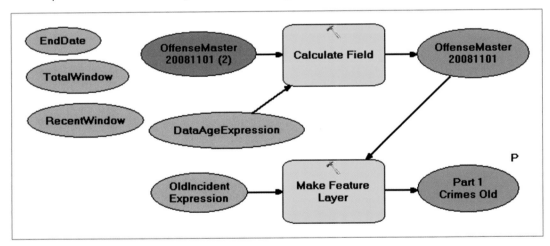

9   Right-click Part 1 Crimes Old, select Properties, and click the Symbology tab. Then click the Import button and go to the SeriousCrimesOld layer in the chapter 10 folder in FinishedExercises and click Add. Click OK. All the work that is done in chapter 4 to symbolize the SeriousCrimesOld layer is stored in this file for reuse.

10   Right-click Part 1 Crimes Old and click Add To Display. Then save your model. ModelBuilder now adds Part 1 Crimes Old to the table of contents when the model is run.

Run partial model

If you have no mistakes, the partial model runs and produces the desired output layer. If you do have mistakes, you get an error message and need to make corrections. Before rerunning the model, remove any model outputs from the table of contents.

1   **From the ModelBuilder Menu bar, click Model > Validate Entire Model.** Note that when you rerun the model in Edit mode, it is critical to perform this step. Validating the model clears Has Been Run elements, designated by a shadow added to a process, so that the model can be run again.

2   **Click Model > Run Entire Model.** If the model runs successfully but no output is added to the table of contents, try right-clicking Part 1 Crimes Old and make sure Add To Display is selected. If not and you turn it on, ModelBuilder adds the output at that time.

3   **From the ArcMap Menu bar, click Bookmarks > Police Sector 1-5.** Clicking this bookmark zooms you to Police Sector 1-5, at the large scale intended for viewing this map.

4   **Save the model, but do not close it.**

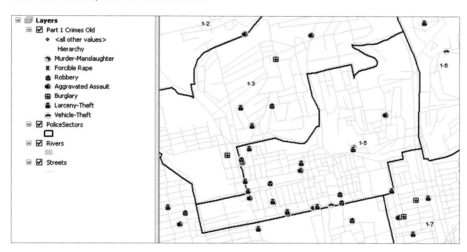

Create a process for Part 1 Crimes Recent, similar to Part 1 Crimes Old. Create a String variable called **NewIncidentExpression** with the following value:

```
"DataAge" >= 0 AND "DataAge" <= (%RecentWindow%-1) AND
"Hierarchy" >= 1 AND "Hierarchy" <= 7
```

For layer symbolization, use the Part 1 Crimes Recent layer file in the chapter 10 folder in FinishedExercises.

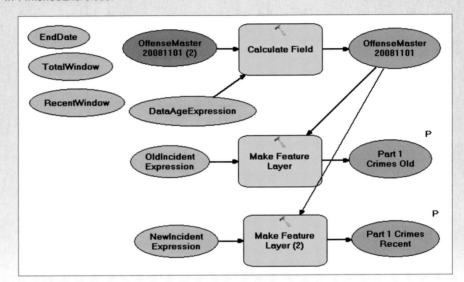

Add parameters and precondition and run the model

1   **In the Produce Field Officers' Pin Map model, make EndDate, TotalWindow, and RecentWindow model parameters.** It is helpful to have ModelBuilder draw the Part 1 Crimes Recent layer first because its symbols are large. Then when Part 1 Crimes Old is drawn second, ArcMap does not cover up its smaller symbols with the larger symbols of Part 1 Crimes Recent. ModelBuilder has a flow-of-processing control, or flow-of-control operator, called "preconditions," that can force elements to be processed in the correct sequence.

2   **Right-click the Make Feature Layer element that creates Part 1 Crimes Old and select Properties. Click the Preconditions tab and select Part 1 Crimes Recent. Click OK.** ModelBuilder draws a dashed line from Part 1 Crimes Recent to the Make Feature Layer of Part 1 Crimes Old, forcing Part 1 Crimes Recent to be drawn first.

3   **Remove Part 1 Crimes Old from the table of contents.**

4   **Save the model. From the model Menu bar, click Model > Validate Entire Model. Then click Model > Run Entire Model.** If you have no errors, the entire model runs smoothly and produces the desired output layers. If you have mistakes, you'll get an error message and have to make corrections. Then if there are any outputs in the table of contents, you'll need to right-click them and click Remove before rerunning the model.

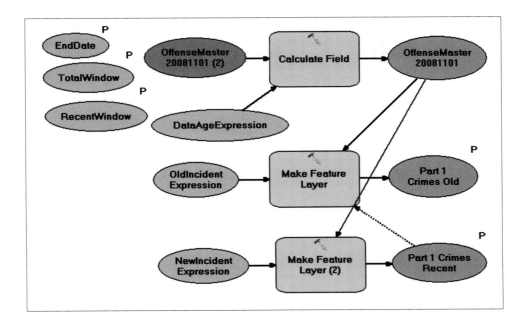

5 After the model runs successfully, remove both output layers from the table of contents.

6 Click Validate the Entire model. Then save and close your model.

7 In ArcToolbox, expand the CrimeMappingTools toolbox. Then right-click Produce Field Officers' Pin Map and click Open. Ignore the warning messages.

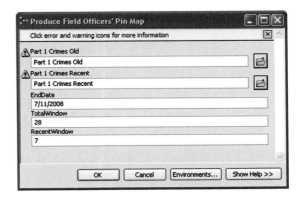

8 Change the end date to **7/31/2008** and click OK.

9 Close the Run window.

10 Save your map document and close ArcMap.

Assignment 10-1

Build a model to produce choropleth maps

In chapter 4, you build choropleth maps for burglaries by police sector for a given month. Such maps are useful for jurisdiction-wide scanning for crime problems, whether it is high levels of crime or large increases in crime.

For this assignment, build a model that automatically produces a burglary map for a month of the user's choice, given aggregate space and time series data that is part of the master file collection. At this stage, you most likely still need a lot of help in structuring and building models, so the assignment statement provides the structure and you need only to implement it. An example of the finished map for July 2008 is shown in the figure.

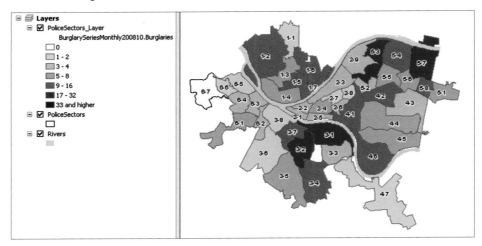

Create new map document

Create a subfolder called **My Toolboxes** and save it to your assignment 10-1 folder in MyAssignments.

Create a new map document called **Assignment10-1YourName.mxd** and save it to your assignment 10-1 folder. Use relative paths for your map document and make the data frame coordinate system state plane for southern Pennsylvania (NAD83 US Feet). Add the following map layers and data:

+ Rivers and PoliceSectors from the Pittsburgh geodatabase in the Data folder

Symbolize Rivers with a blue color fill and PoliceSectors as hollow with a label of **Sector**. Your model will create another layer from PoliceSectors that has polygons symbolized based on burglary levels using a layer file created in chapter 3. That layer will be drawn over the top of the hollow-fill layer, but the hollow-fill layer will fill in the polygon boundaries for police sectors with zero burglaries and its labels will be visible for all police sectors.

Set environment and create new toolbox and model

Set Geoprocessing Options to the settings used in the exercises in this chapter. Use the following steps:

1   Set the environments in ArcToolbox by setting Current Workspace to your assignment 10-1 folder and Scratch Workspace to the Temp geodatabase in the Scratch folder.

2   Create a new toolbox called **Assignment10-1** and save it to your My Toolboxes folder. Create a new model in that toolbox called **Choropleth Map**. Right-click the new model and select Properties. Click the General tab, fill out the documentation, and select the "Store relative path names" check box.

Build model

The model you need to build is shown in the figure. Suggestions and comments follow, keyed to the numbers in the model diagram.

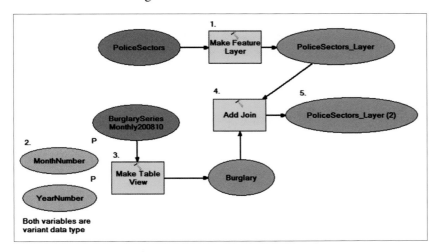

1.   This process uses the Make Feature Layer tool from the Layers and Table Views toolset in the Data Management toolbox. This tool creates a temporary feature layer based on the PoliceSectors input. The tool does not require a query expression, because you need only a full copy of the input shapefile in the feature layer and not a portion of it. The reason you need the temporary layer is because it will not have any previous joins, since it is freshly created, and will thus allow you to join selected data to it for display. If you were to join the data to PoliceSectors directly, you would have to remove the previous join to rerun the model, and that step would be tricky.

2.   Create two variables here, both with the Variant data type. Give **MonthNumber** the default value 7 and **YearNumber** the default value **2008**. Set both variables as parameters.

3.   Make Table View is another tool in the Layers and Table Views toolset. The input is BurglarySeriesMonthly200810, a table in the chapter 9 geodatabase in FinishedExercises, and it has data to produce 10 monthly maps, from January through October 2008. Build an expression using the SQL Query Builder for a month and a year. Use specific values from the Get Unique Values button for Month and Year. Then after the query is built, replace the specific values with the names of your parameter variables by using in-line substitution.

4. The Add Join tool is in the Joins toolset in the Data Management toolbox. Join both the Burglary table and the feature layer by using the Sector attribute, which is in both inputs.

5. Next, simply add a layer file to symbolize the output of the Add Join process. Use the CrimeChoropleth2 layer in the chapter 4 folder in FinishedExercises.

What to turn in

Use a compression program to compress and save your assignment 10-1 folder to **Assignment10-1YourName.zip**. Turn in your compressed file.

Assignment 10-2

Build a model to produce size-graduated point marker maps

In chapter 4, you build size-graduated point marker maps for burglaries by police sector. These maps are useful for targeting patrols, especially in the vicinity of repeat burglaries.

For this assignment, build a model that automatically produces these maps from the offense master feature layer for a crime type and date interval of the user's choosing. An example of the finished map for 28 days, ending July 15, 2008, is shown in the figure.

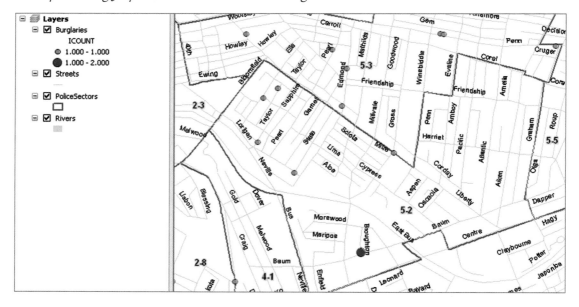

Create folders and a new map document

Create a subfolder called **My Toolboxes** and save it to your assignment 10-2 folder in MyAssignments.

Create a new map document called **Assignment10-2YourName.mxd** and save it to your assignment 10-2 folder. Use relative paths for your map document and make the data frame coordinate system state plane for southern Pennsylvania (NAD83 US Feet). Add the following map layers and data:

+ Rivers, Streets, and PoliceSectors from the Pittsburgh geodatabase in the Data folder

Symbolize Rivers with a blue color fill, PoliceSectors as hollow with a label of **Sector**, and Streets with a gray color and labeled by name.

Set environment and create a new toolbox and model

Set Geoprocessing Options to the settings used in the exercises in this chapter. Use the following steps:

1   Set the environments in ArcToolbox by setting Current Workspace to your assignment 10-2 folder and Scratch Workspace to the Temp geodatabase in the Scratch folder.

2   Create a new toolbox called **Assignment10-2** and save it to your My Toolboxes folder. Create a new model in that toolbox called **Graduated Point Markers**. Right-click your new model and select Properties. Click the General tab, fill out the documentation, and select the "Store relative path names" check box.

Build model

The model you need to build is shown in the figure. Useful suggestions and comments follow, keyed to the numbers in the model diagram.

1.   The Calculate Field process is similar to the process of the same name in the exercises in this chapter for creating the field officers' pin map. Refer to the exercises. The input is the Offenses2008 feature class in the Police geodatabase in the Data folder. The needed date attribute is [DateOccur].

2.   Create three Variant data type variables, **EndDate**, **WindowLength**, and **HierarchyNum**, and set them as parameters. Give EndDate a default value of **7/15/2008**, WindowLength a default value of **28**, and Hierarchy a default value of **5** (the code for burglaries).

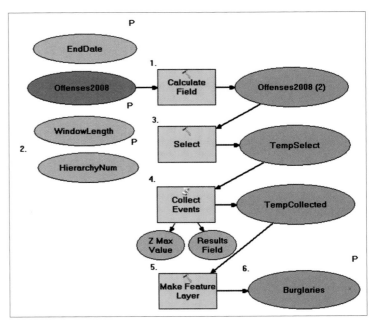

3. The Select tool is in the Extract toolset in the Analysis toolbox. Start building the expression with the SQL Query Builder, using the Get Unique Values button to yield `"Age" >= 0 AND "Age" <= 27 AND "Hierarchy" = 5`. Then use in-line substitution to replace the values 27 and 5 with values provided by the user in the parameters WindowLength and HierarchyNum. The arithmetic requires you to subtract 1 from WindowLength to get the correct output. The needed expression, all on one line, is

    `"Age" >= 0 AND "Age" <= (%WindowLength% - 1) AND "Hierarchy" = %HierarchyNum%`

4. Save the output feature class to your scratch folder.

5. Collect Events is a tool you use in chapter 3. Place its output in your scratch folder. By using the same name, **TempCollected**, in repeated runs of the model, you can avoid accumulating a lot of stored feature classes. The options you set make ModelBuilder delete earlier copies of these temporary files when creating new ones.

6. The Make Feature Layer tool merely creates a temporary layer file (Burglaries), which can be set as a parameter for the user to name. Set Burglaries as a parameter and add it to the display. Use the GraduatedMarkers layer in the chapter 3 folder in FinishedExercises to symbolize Burglaries. Be sure to add the new layer to the display.

What to turn in

Use a compression program to compress and save your assignment 10-2 folder to **Assignment10-2YourName.zip**. Turn in your compressed file.

Appendix A

Tools and buttons

ArcGIS Animation toolbar

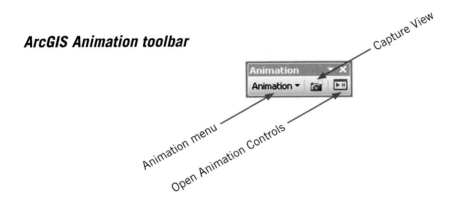

ArcGIS Draw toolbar

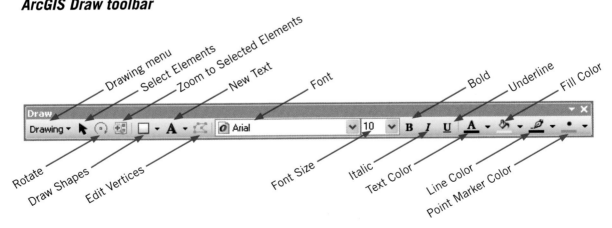

ArcGIS Editor toolbar

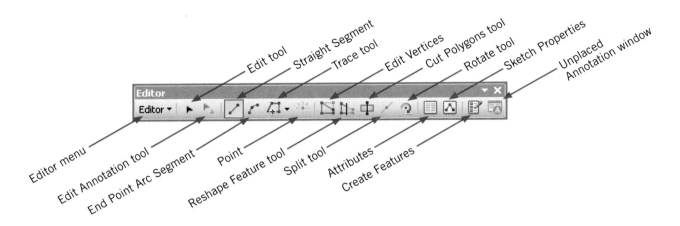

ArcGIS Geocoding toolbar

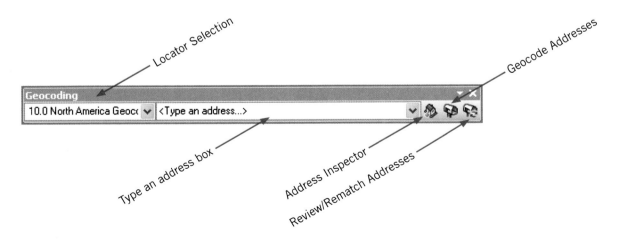

ArcGIS Standard toolbar

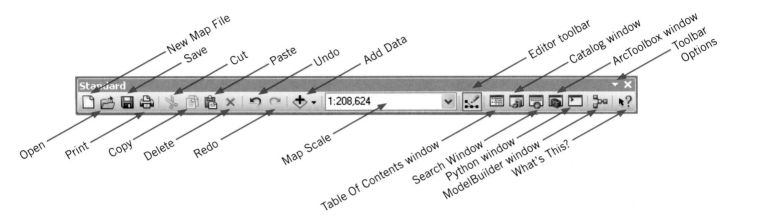

ArcGIS Tools toolbar

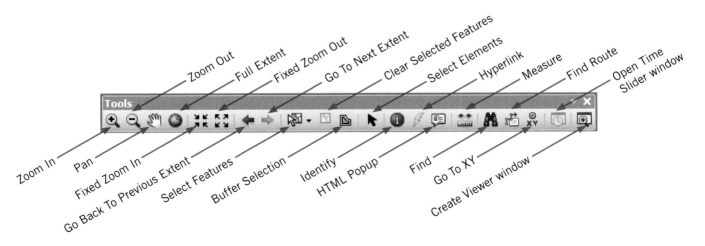

Appendix B

Task index

Task or concept: software tool, **example tutorial(s) in which it appears**

Aggregate data: in ArcMap, **7-3**, **9-4**
Aggregate data: spatial joins, **7-3**, **9-3**
Animation: add group layer, **6-2**
Animation: Animation Manager, **6-1**, **6-2**
Animation: Animation toolbar, **6-1**, **6-2**
Animation: create date stamp in Microsoft Office Excel, **6-1**
Animation: create time animation, **6-1**
Animation: export to video, **6-1**
Animation: play, **6-1**, **6-2**
Animation: save, **6-1**
Animation: set time properties, **6-1**
Animation: Time tab in Animation Manager, **6-1**
ArcCatalog: Auto Hide, **2-2**
ArcCatalog: expand Catalog tree, **2-2**
ArcCatalog: preview layers in, **2-2**
ArcCatalog: utilities, **2-2**
ArcToolbox: add toolbox, **10-1**
ArcToolbox: environment, **7-2**
ArcToolbox: overview, **7-2**
Attribute table: add a field, **8-3**
Attribute table: add data, **8-5**, **9-3**
Attribute table: assign aliases to attributes, **8-4**
Attribute table: calculate centroids for polygons, **8-4**
Attribute table: change attribute data type, **8-3**
Attribute table: create new attribute, **6-2**
Attribute table: delete field, **8-2**

Attribute table: field statistics, **3-3**
Attribute table: Identify tool, **3-2**
Attribute table: join table to map, **4-2, 7-3, 8-3**
Attribute table: open, **2-1**
Attribute table: set field properties, **3-2, 4-1**
Attribute table: sort a field, **3-3**
Attribute table: sort records, **3-3**

Definition query: AND connector, **5-1**
Definition query: crime type criteria, **5-1**
Definition query: date range criteria, **5-1**
Definition query: OR connector, **5-1**
Definition query: overview, **2-3, 4-1**
Definition query: person attribute criteria, **5-1**
Definition query: save file (.exp), **5-1**
Definition query: time of day criteria, **5-1**
Digitize: line features, **8-5**
Digitize: polygon features, **8-5**
Digitize: Snapping tools, **8-5**
Digitize: Trace tool, **8-5**
Download: Census TIGER/Line data (Esri), **8-1**
Download: preprocess downloaded census data in Excel, **8-1**
Download: U.S. Census Cartographic Boundary Files, **8-1**
Download: U.S. Census data, **8-1**

Edit: edit streets, **9-2**
Edit: Editor toolbar, **8-5**

Geocode: add address locators, **9-1**
Geocode: address match statistics, **9-2**
Geocode: address rematch, **9-2**
Geocode: by streets, **9-1, 9-3**
Geocode: rebuild street locator, **9-2**
Geocode: rematch addresses automatically, **9-2**
Geocode: rematch interactively by clicking the map, **9-2**
Geocode: rematch interactively by correcting input addresses, **9-2**
Geocode: use alias tables, **9-2**
Geoprocessing: append map layer, **9-3, 9-4**
Geoprocessing: clean address data in Excel, **9-2**
Geoprocessing: clip features, **8-2**
Geoprocessing: dissolve polygons, **8-4**
Geoprocessing: merge polygons, **8-4**

Labels: halo, **4-2**
Labels: label layers, **4-1**
Layout: add elements, **4-3**
Layout: add legend, **4-3**
Layout: add neatline, **4-3**
Layout: add note, **4-3**
Layout: add scale bar, **4-3**
Layout: add title, **4-3**
Layout: change page orientation, **4-3**
Layout: export, **2-4**, **4-3**
Layout: Layout View, **2-4**

Map design: choropleth map, **3-2**, **4-2**
Map design: create change data for choropleth map in Excel, **4-2**
Map design: early warning, **3-2**
Map design: pin map, **3-3**
Map design: protecting privacy in crime data, **3-1**
Map design: size-graduated point markers, **7-1**
Map document: add XY layer, **6-1**
Map document: create new, **4-1**, **8-2**
Map document: open existing, **2-1**
Map document: properties, **2-1**, **4-1**
Map document: relative paths, **4-1**
Map layer: add centroid coordinates to polygon layer, **7-3**
Map layer: add spatial data, **8-1**
Map layer: create new feature class, **8-5**
Map layer: export feature class, **7-3**
Map layer: file geodatabase, **2-1**, **2-2**
Map layer: properties, **2-3**
Map layer: save a layer file, **4-1**
Miscellaneous: draw shapes, **3-2**, **6-2**, **7-1**
Miscellaneous: find features, **3-1**
Miscellaneous: read feature coordinates, **2-1**
Miscellaneous: Undo button, **2-1**
ModelBuilder: add labels for documentation, **10-2**
ModelBuilder: add model name and description, **10-3**
ModelBuilder: append process, **10-2**
ModelBuilder: calculate field process, **10-2**
ModelBuilder: copy process, **10-2**
ModelBuilder: create new model, **10-2**, **10-3**
ModelBuilder: create variables, **10-2**
ModelBuilder: Edit mode, **10-1**
ModelBuilder: fix bugs, **10-1**
ModelBuilder: geocoding process, **10-2**

ModelBuilder: make feature layer process, **10-3**
ModelBuilder: parameters, **10-2**
ModelBuilder: precondition, **10-2**
ModelBuilder: run a model, **10-1**, **10-2**
ModelBuilder: set parameters, **10-2**
ModelBuilder: set process properties, **10-2**, **10-3**
ModelBuilder: set properties, **10-2**
ModelBuilder: spatial join process, **10-2**

Navigation: create spatial bookmark, **3-1**
Navigation: hyperlinks, **3-3**, **4-1**
Navigation: Magnifier window, **3-2**
Navigation: measure distances, **3-2**
Navigation: Overview window, **3-2**
Navigation: pan, **3-1**
Navigation: zoom in, **2-1**, **3-1**
Navigation: zoom to full extent, **3-1**
Navigation: zoom to next extent, **3-1**
Navigation: zoom to previous extent, **3-1**
Navigation: zoom to selected features, **3-3**

Select: by attributes, **5-1**
Select: by location, **8-2**
Select: clear selected features, **3-3**
Select: extract features by Select, **8-2**
Select: features, **3-3**
Select: records, **3-3**
Select: set selectable layers, **3-2**, **4-1**
Select: set selection color, **3-3**
Select: show selected records, 5-1
Spatial Analyst extension: create contours from raster map, **7-2**
Spatial Analyst extension: kernel density smoothing, **7-2**
Spatial Analyst extension: overview, **7-2**
Spatial query: buffer statistics, **5-2**
Spatial query: create buffer, **5-2**
Spatial query: create multiple-ring buffer, **5-2**
Spatial Statistics: Average Nearest Neighbor index, **7-1**
Spatial Statistics: Getis-Ord Gi* test, **7-3**
Spatial Statistics: toolbox, **7-1**
Symbolize: change, **2-1**, **2-2**
Symbolize: create custom numeric scale, **4-2**
Symbolize: create size-graduated point markers, **4-2**, **7-1**

Symbolize: crime point markers, **4-1**
Symbolize: density surface, **7-2**
Symbolize: layer properties, **2-2**
Symbolize: size-graduated point markers, **4-2**
Symbolize: unique values, **4-1**

Table of contents: add a layer, **2-1**, **4-1**
Table of contents: add a layer file, **4-1**
Table of contents: add aerial photo, **2-1**
Table of contents: add map from ArcGIS Online, **2-1**
Table of contents: change layer display order, **3-2**
Table of contents: change layer name, **3-2**
Table of contents: remove a layer, **2-1**
Table of contents: rename layer, **2-2**
Table of contents: set projection for data frame, **8-2**
Table of contents: set visible scale range, **3-2**, **4-1**, **4-2**
Table of contents: turn a layer off, **2-1**
Table of contents: turn a layer on, **2-1**
Table of contents: turn labels off, **2-1**

Appendix C

Handling data
and homework files

Knowing how to properly install and save the *GIS Tutorial for Crime Analysis* data files is an important part of learning GIS. The Maps and Data DVD in the back of the book has GIS map layers for Pittsburgh, Pennsylvania, including street centerlines, rivers, census tracts, police sectors, neighborhoods, all 42,000 crime offenses for 2008 mapped as points, and many more map layers and datasets. There are nearly 2,000 files in more than 120 folders, amounting to 400 MB of data.

Once you install the data (please see appendix F), you can store the files in any folder on any letter drive (we recommend C) and use relative paths in all map documents you create (see tutorial 4-1 in chapter 4 for instructions).

If you are working in a public computer lab, the data files will very likely be deleted every night, so be sure to save a backup copy of the files to another disk or a thumb drive.

Saving homework files

The MyAssignments folder is where you will store *all* assignment solutions you create. It is important to preserve the following path, C:\EsriPress\GISTCrime\MyAssignments, for your solutions to work correctly. There will be a variety of files to submit for each assignment. Please follow the ensuing instructions carefully.

Homework file example

Let's say, for example, that you are doing the first assignment for chapter 4, and there are five files to submit. The files required for assignment 4-1 include the following:

- Assignment4-1YourName.mxd. This is a map document that stores instructions for displaying a map composition for use by others.

- Assignment4-1YourName.docx. This is a Word document that includes map images exported from Assignment4-1YourName.mxd.

- Map4-1a.YourName.jpg, Map4-1bYourName.jpg, and Map4-1cYourName.jpg. These are the exported map images.

You will save these files to your assignment 4-1 folder, which is in the chapter 4 folder in MyAssignments, and then compress and save your assignment 4-1 folder to Assignment4-1YourName.zip. Turn in your compressed file.

Note: The map documents you create will "point to" map files and data files that already exist in the EsriPress\GISTCrime folders on your computer. Your new map documents will point to existing data files in the Data and FinishedExercises folders, just as they will point to the data files you create and store in your MyExercises and MyAssignments folders. ArcGIS does not make redundant copies of these files in its map document (.mxd) files. You will not turn in these existing maps as parts of solutions for assignments. Instead, the instructor will use his or her own copy of the needed files, using identical paths. The instructor will then plug in the new components that you create and submit as assignment solutions. This saves a lot of time and file space—map files are very large. Please be sure to ask questions if you have trouble understanding what you need to submit.

Appendix D

Data source credits

Chapter 1 data sources
Diagrams courtesy of Wilpen L. Gorr

Chapter 2 data sources
\EsriPress\GISTCrime\Data\Pittsburgh.gdb\Streets, from Esri Census 2000 TIGER/Line Data.

\EsriPress\GISTCrime\Data\Pittsburgh.gdb\Rivers, Blocks, Tracts, courtesy of the U.S. Census Bureau TIGER.

\EsriPress\GISTCrime\Data\Pittsburgh.gdb\ZoningCommercialBuffer, courtesy of Allegheny County, Chief Information Officer and Wilpen L. Gorr.

Screen captures of image mosaic, courtesy of i-cubed.

\EsriPress\GISTCrime\Data\Police.gdb\Offenses2008, courtesy of the Pittsburgh Bureau of Police.

Chapter 3 data sources
\EsriPress\GISTCrime\Data\Pittsburgh.gdb\Streets, from Esri Census 2000 TIGER/Line Data.

\EsriPress\GISTCrime\Data\Pittsburgh.gdb\Rivers, courtesy of the U.S. Census Bureau TIGER.

\EsriPress\GISTCrime\Data\Pittsburgh.gdb\Neighborhoods, courtesy of the U.S. Census Bureau TIGER and Wilpen L. Gorr.

\EsriPress\GISTCrime\Data\Pittsburgh.gdb\PoliceSectors, courtesy of the Pittsburgh Bureau of Police.

\EsriPress\GISTCrime\MyExercises\FinishedExercises\Chapter9\Chapter9.gdb\PropertyOffensesJuly2008Protected, BurglaryJuly2008Graduated, courtesy of the Pittsburgh Bureau of Police and Wilpen L. Gorr.

Chapter 4 data sources
\EsriPress\GISTCrime\Data\Pittsburgh.gdb\Rivers, courtesy of the U.S. Census Bureau TIGER.

\EsriPress\GISTCrime\Data\Pittsburgh.gdb\Streets, from Esri Census 2000 TIGER/Line Data.

\EsriPress\GISTCrime\Data\Pittsburgh.gdb\PoliceSectors, courtesy of the Pittsburgh Bureau of Police.

\EsriPress\GISTCrime\Data\Police.gdb\Offenses2008, courtesy of the Pittsburgh Bureau of Police.

BurglaryMonthlySeriesComplete200810.txt, BurglaryMonthlySeriesComplete200810.txt, Chapter9.gdb\OffenseMaster200811101, and Chapter9.gdb\PropertyOffensesJuly2008Protected, courtesy of the Pittsburgh Bureau of Police and Wilpen L. Gorr.

\EsriPress\GISTCrime\Data\CodeTables\PittsburghPoliceOffenseCodes.xls, courtesy of the Pittsburgh Bureau of Police.

\EsriPress\GISTCrime\Data\RawData\Gun20080221985.jpg, from Shutterstock, courtesy of Billy Hoiler.

Chapter 5 data sources
\EsriPress\GISTCrime\Data\Pittsburgh.gdb\Rivers, Tracts, courtesy of the U.S. Census Bureau TIGER.

\EsriPress\GISTCrime\Data\Pittsburgh.gdb\Streets, from Esri Census 2000 TIGER/Line Data.

\EsriPress\GISTCrime\Data\Pittsburgh.gdb\Neighborhoods, courtesy of the U.S. Census Bureau TIGER and Wilpen L. Gorr.

\EsriPress\GISTCrime\Data\Pittsburgh.gdb\PoliceSectors, courtesy of the Pittsburgh Bureau of Police.

\EsriPress\GISTCrime\Data\Police.gdb\Offenses2008, courtesy of the Pittsburgh Bureau of Police.

\EsriPress\GISTCrime\Data\Pittsburgh.gdb\ZoningCommercialBuffer, courtesy of Allegheny County, Chief Information Officer and Wilpen L. Gorr.

\EsriPress\GISTCrime\Data\Pittsburgh.gdb\CheckCashingPlaces, BarsSouthside, courtesy of Wilpen L. Gorr.

\EsriPress\GISTCrime\Data\Downloads\CensusBureau\Poverty.xls, courtesy of the U.S. Census Bureau TIGER.

Chapter 6 data sources

\EsriPress\GISTCrime\Data\Pittsburgh.gdb\Rivers, Tracts courtesy of the U.S. Census Bureau TIGER.

\EsriPress\GISTCrime\Data\Pittsburgh.gdb\Streets, from Esri Census 2000 TIGER/Line Data.

\EsriPress\GISTCrime\Data\Pittsburgh.gdb\PoliceSectors, courtesy of the Pittsburgh Bureau of Police.

\EsriPress\GISTCrime\MyExercises\FinishedExercises\Chapter5\Chapter5.gdb\Offenses2008, courtesy of the Pittsburgh Bureau of Police.

\EsriPress\GISTCrime\Data\Police.gdb\ATMRobberies, ATMCumulative, courtesy of the Pittsburgh Bureau of Police.

\EsriPress\GISTCrime\Data\Police.gdb\CADDrugsSummer08, courtesy of the Allegheny County Department of Emergency Services.

Chapter 7 data sources

\EsriPress\GISTCrime\Data\Pittsburgh.gdb\Rivers, Blocks, courtesy of the U.S. Census Bureau TIGER.

\EsriPress\GISTCrime\Data\Pittsburgh.gdb\Streets, from Esri Census 2000 TIGER/Line Data.

\EsriPress\GISTCrime\Data\Pittsburgh.gdb\PoliceSectors, courtesy of the Pittsburgh Bureau of Police.

\EsriPress\GISTCrime\Data\Police.gdb\CADDrugsSummer08, courtesy of the Allegheny County Department of Emergency Services.

\EsriPress\GISTCrime\MyExercises\FinishedExercises\Chapter7\Chapter7.gdb\SuppliedHotSpots, DrugsCollected, HotSpotContour1000, courtesy of Wilpen L. Gorr.

Chapter 8 data sources

\EsriPress\GISTCrime\MyExercises\FinishedExercises\Chapter8\Downloads\CensusBureau\ tl_2009_42003_areawater.shp, tl_2009_42003_cousub00.shp, tl_2009_42003_tabblock00.shp, tl_2009_42003_tract00.shp, dc_dec_2000_sf3_u_data1.txt, dc_dec_2000_sf3_u_geo.txt, courtesy of the U.S. Census Bureau TIGER.

\EsriPress\GISTCrime\Data\Pittsburgh.gdb\ZoningCommercialBuffer, courtesy of Allegheny County, Chief Information Officer and Wilpen L. Gorr.

Chapter 9 data sources

\EsriPress\GISTCrime\Data\Pittsburgh.gdb\Rivers, Tracts, courtesy of the U.S. Census Bureau TIGER.

\EsriPress\GISTCrime\Data\Pittsburgh.gdb\Streets, from Esri Census 2000 TIGER/Line Data.

\EsriPress\GISTCrime\Data\Pittsburgh.gdb\PoliceSectors, courtesy of the Pittsburgh Bureau of Police.

\EsriPress\GISTCrime\MyExercises\Chapter9\Chapter9.gdb\Blocks, from Esri Census 2000 TIGER/Line Data.

\EsriPress\GISTCrime\Data\RawData\OffenseMaster20081031.xls, OffenseUpdate20081101.xls, courtesy of the Pittsburgh Police Bureau and Wilpen L. Gorr.

\EsriPress\GISTCrime\Data\Locators\PittsburghAlias.csv, courtesy of Wilpen L. Gorr.

\EsriPress\GISTCrime\Data\Police.gdb\Offenses2008, courtesy of the Pittsburgh Bureau of Police.

EsriPress\GISTCrime\Data\RawData\CADWeapons20081231.xls, CADDrugs20081231.xls, courtesy of the Allegheny County Department of Emergency Services.

Chapter 10 data sources

\EsriPress\GISTCrime\Data\Pittsburgh.gdb\Rivers, courtesy of the U.S. Census Bureau TIGER.

\EsriPress\GISTCrime\Data\Pittsburgh.gdb\Streets, from Esri Census 2000 TIGER/Line data.

\EsriPress\GISTCrime\Data\Pittsburgh.gdb\Neighborhood, courtesy of the U.S. Census Bureau TIGER/Line data and Wilpen L. Gorr.

\EsriPress\GISTCrime\Data\Police.gdb\Offenses2008, July2008Protected, courtesy of the Pittsburgh Bureau of Police and Wilpen L. Gorr.

\EsriPress\GISTCrime\Data\Pittsburgh.gdb\PoliceSectors, courtesy of the Pittsburgh Bureau of Police.

\EsriPress\GISTCrime\Data\RawData\OffenseMaster20081031.xlsx, OffenseUpdate20081101.xlsx, courtesy of the Pittsburgh Bureau of Police and Wilpen L. Gorr.

\EsriPress\GISTCrime\Data\Locators\PittsburghAlias.csv, courtesy of Wilpen L. Gorr.

EsriPress\GISTCrime\MyExercises\FinishedExercises\Chapter8\Chapter8.gdb\BurglarySeriesMonthly200810, courtesy of the Pittsburgh Bureau of Police and Wilpen L. Gorr.

Appendix E

Data license agreement

Important:
Read carefully before opening the sealed media package

ENVIRONMENTAL SYSTEMS RESEARCH INSTITUTE, INC. (ESRI), IS WILLING TO LICENSE THE ENCLOSED DATA AND RELATED MATERIALS TO YOU ONLY UPON THE CONDITION THAT YOU ACCEPT ALL OF THE TERMS AND CONDITIONS CONTAINED IN THIS LICENSE AGREEMENT. PLEASE READ THE TERMS AND CONDITIONS CAREFULLY BEFORE OPENING THE SEALED MEDIA PACKAGE. BY OPENING THE SEALED MEDIA PACKAGE, YOU ARE INDICATING YOUR ACCEPTANCE OF THE ESRI LICENSE AGREEMENT. IF YOU DO NOT AGREE TO THE TERMS AND CONDITIONS AS STATED, THEN ESRI IS UNWILLING TO LICENSE THE DATA AND RELATED MATERIALS TO YOU. IN SUCH EVENT, YOU SHOULD RETURN THE MEDIA PACKAGE WITH THE SEAL UNBROKEN AND ALL OTHER COMPONENTS TO ESRI.

Esri license agreement

This is a license agreement, and not an agreement for sale, between you (Licensee) and Environmental Systems Research Institute, Inc. (Esri). This Esri License Agreement (Agreement) gives Licensee certain limited rights to use the data and related materials (Data and Related Materials). All rights not specifically granted in this Agreement are reserved to Esri and its Licensors.

Reservation of Ownership and Grant of License: Esri and its Licensors retain exclusive rights, title, and ownership to the copy of the Data and Related Materials licensed under this Agreement and, hereby, grant to Licensee a personal, nonexclusive, nontransferable, royalty-free, worldwide license to use the Data and Related Materials based on the terms and conditions of this Agreement. Licensee agrees to use reasonable effort to protect the Data and Related Materials from unauthorized use, reproduction, distribution, or publication.

Proprietary Rights and Copyright: Licensee acknowledges that the Data and Related Materials are proprietary and confidential property of Esri and its Licensors and are protected by United States copyright laws and applicable international copyright treaties and/or conventions.

Permitted Uses: Licensee may install the Data and Related Materials onto permanent storage device(s) for Licensee's own internal use.

Licensee may make only one (1) copy of the original Data and Related Materials for archival purposes during the term of this Agreement unless the right to make additional copies is granted to Licensee in writing by Esri.

Licensee may internally use the Data and Related Materials provided by Esri for the stated purpose of GIS training and education.

Uses Not Permitted: Licensee shall not sell, rent, lease, sublicense, lend, assign, time-share, or transfer, in whole or in part, or provide unlicensed Third Parties access to the Data and Related Materials or portions of the Data and Related Materials, any updates, or Licensee's rights under this Agreement.

Licensee shall not remove or obscure any copyright or trademark notices of Esri or its Licensors.

Term and Termination: The license granted to Licensee by this Agreement shall commence upon the acceptance of this Agreement and shall continue until such time that Licensee elects in writing to discontinue use of the Data or Related Materials and terminates this Agreement. The Agreement shall automatically terminate without notice if Licensee fails to comply with any provision of this Agreement. Licensee shall then return to Esri the Data and Related Materials. The parties hereby agree that all provisions that operate to protect the rights of Esri and its Licensors shall remain in force should breach occur.

Disclaimer of Warranty: The Data and Related Materials contained herein are provided "as-is," without warranty of any kind, either express or implied, including, but not limited to, the implied warranties of merchantability, fitness for a particular purpose, or noninfringement. Esri does not warrant that the Data and Related Materials will meet Licensee's needs or expectations, that the use of the Data and Related Materials will be uninterrupted, or that all nonconformities, defects, or errors can or will be corrected. Esri is not inviting reliance on the Data or Related Materials for commercial planning or analysis purposes, and Licensee should always check actual data.

Data Disclaimer: The Data used herein has been derived from actual spatial or tabular information. In some cases, Esri has manipulated and applied certain assumptions, analyses, and opinions to the Data solely for educational training purposes. Assumptions, analyses, opinions applied, and actual outcomes may vary. Again, Esri is not inviting reliance on this Data, and the Licensee should always verify actual Data and exercise their own professional judgment when interpreting any outcomes.

Limitation of Liability: Esri shall not be liable for direct, indirect, special, incidental, or consequential damages related to Licensee's use of the Data and Related Materials, even if Esri is advised of the possibility of such damage.

No Implied Waivers: No failure or delay by Esri or its Licensors in enforcing any right or remedy under this Agreement shall be construed as a waiver of any future or other exercise of such right or remedy by Esri or its Licensors.

Order for Precedence: Any conflict between the terms of this Agreement and any FAR, DFAR, purchase order, or other terms shall be resolved in favor of the terms expressed in this Agreement, subject to the government's minimum rights unless agreed otherwise.

Export Regulation: Licensee acknowledges that this Agreement and the performance thereof are subject to compliance with any and all applicable United States laws, regulations, or orders relating to the export of data thereto. Licensee agrees to comply with all laws, regulations, and orders of the United States in regard to any export of such technical data.

Severability: If any provision(s) of this Agreement shall be held to be invalid, illegal, or unenforceable by a court or other tribunal of competent jurisdiction, the validity, legality, and enforceability of the remaining provisions shall not in any way be affected or impaired thereby.

Governing Law: This Agreement, entered into in the County of San Bernardino, shall be construed and enforced in accordance with and be governed by the laws of the United States of America and the State of California without reference to conflict of laws principles. The parties hereby consent to the personal jurisdiction of the courts of this county and waive their rights to change venue.

Entire Agreement: The parties agree that this Agreement constitutes the sole and entire agreement of the parties as to the matter set forth herein and supersedes any previous agreements, understandings, and arrangements between the parties relating hereto.

Appendix F

Installing the data and software

GIS *Tutorial for Crime Analysis* includes one DVD with exercise data. A trial version of ArcGIS Desktop 10 (single use) software can be downloaded at **http://www.esri.com/180daytrial** with instructions to follow.

Installation of the exercise data DVD takes about five minutes and requires 380 megabytes of hard disk space.

Installation of the software requires at least 2.4 gigabytes of hard disk space (more if you choose to load the optional extension products). Installation times will vary with your computer's speed and available memory.

If you already have a licensed copy of ArcGIS Desktop 10 installed on your computer (or accessible through a network), do not install the software. Use your licensed software to do the exercises in this book. If you have an older version of ArcGIS installed on your computer, you must uninstall it before you can install the software from the download site.

The exercises in this book work with ArcGIS 10. Using another software version is not recommended.

Installing the exercise data

Follow the steps below to install the exercise data. Do not copy the files directly from the DVD to your hard drive. A direct file copy does not remove write-protection from the files, and this causes data editing exercises not to work. In addition, a direct file copy will not enable the automatic uninstall feature.

1   Put the data DVD in your computer's DVD drive. A start-up screen will appear. If your auto-run is disabled, navigate to the contents of your DVD drive and double-click the Esri.exe file to begin.

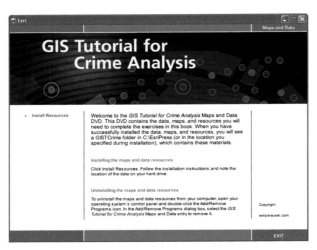

2   Read the welcome screen, and then click the Install exercise data link. This starts the Setup wizard.

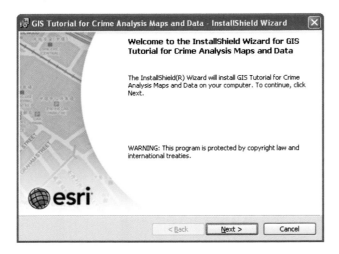

3   Click Next. Read and accept the license agreement terms, and then click next.

4 Accept the default installation folder or click Browse and navigate to the drive or folder location where you want to install the data. If you choose an alternate location, please make note of it as the book's exercises direct you to **C:\EsriPress**.

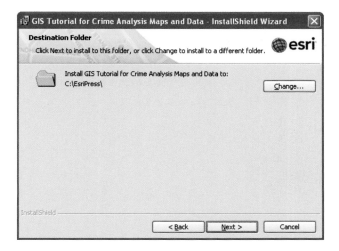

5 Click Next. The installation will take some time. When the installation is finished, you see the following message:

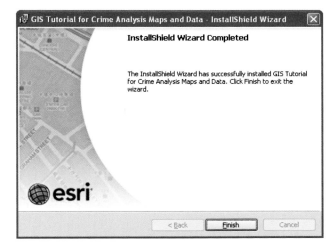

6 Click Finish. The exercise data is installed on your computer in a folder called GISTCrime.

If you have a licensed copy of ArcGIS Desktop 10 installed on your computer, you are ready to start *GIS Tutorial for Crime Analysis*. Otherwise, follow the "Installing the software" instructions to install and register the software.

Uninstalling the exercise data

To uninstall the exercise data from your computer, open your operating system's control panel and double-click the Add/Remove Programs icon. In the Add/Remove Programs dialog box, select the following entry and follow the prompts to remove it:

- GIS Tutorial for Crime Analysis Maps and Data

Installing the software

A 180-day trial version of ArcGIS Desktop 10 (single use) software can be downloaded at `http://www.esri.com/180daytrial`. Use the code printed on the inside back cover of this book to access the download site, and follow the on-screen instructions to download and register the software. **Note:** the code can only be used once.